AS Level Maths is Really Hard

AS level maths is seriously tricky — no question about that.

We've done everything we can to make things easier for you. We've scrutinised past paper questions and we've gone through the syllabuses with a fine-toothed comb. So we've found out exactly what you need to know, then explained it simply and clearly.

We've stuck in as many helpful hints as you could possibly want — then we even tried to put some funny bits in to keep you awake.

We've done our bit — the rest is up to you.

What CGP's All About

The central aim of Coordination Group Publications is to produce top quality books that are carefully written, immaculately presented and astonishingly witty — whilst always making sure they exactly cover the syllabus for each subject.

And then we supply them to as many people as we possibly can, as <u>cheaply</u> as we possibly can.

Contents

Section One — Data

Section Two — Probability

Section Three — Probability Distributions

Section Four — Correlation and Regression

Section Five — Hypothesis Testing

This book covers all the major topics for Edexcel, OCR A, OCR MEI, AQA A and AQA B boards. There are notes at the top of some pages to tell you if there's a bit you can ignore.

Published by Coordination Group Publications Ltd

Typesetters:
Martin Chester, Sharon Watson
Contributors:
Dave Harding, Claire Jackson, Garry Rowlands, Chris Worth
Editors:
Charley Darbishire, Simon Little, Tim Major and Andy Park
Many thanks to Glenn Rogers *for proofreading.*

ISBN 1 84146 985 8
Groovy website: www.cgpbooks.co.uk
Jolly bits of clipart from CorelDRAW
Printed by Elanders Hindson, Newcastle upon Tyne.

Histograms

Skip these two pages if you're doing AQA A S1.

Histograms are glorified bar charts. The main difference is that you plot the frequency density (rather than the frequency). Frequency density is easy to find — you just divide the frequency by the width of a class.

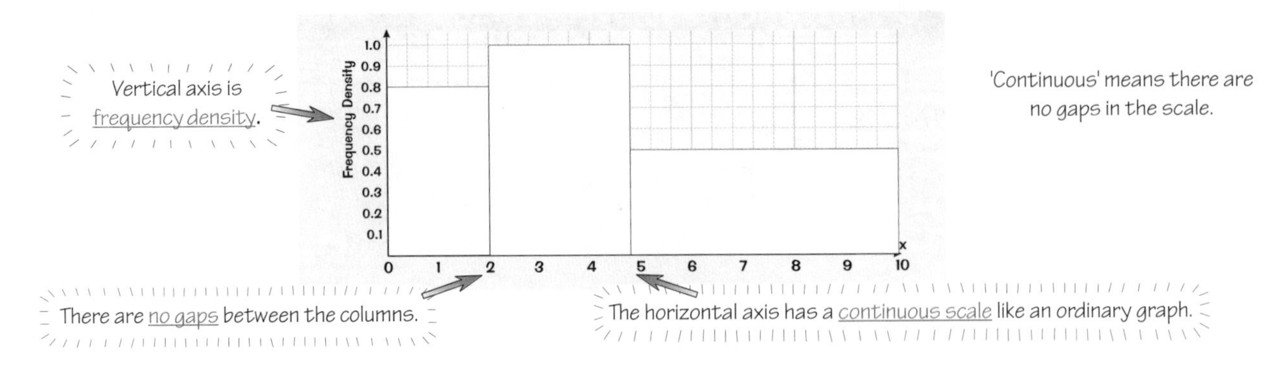

To Draw a Histogram it's best to Draw a Table First

Getting histograms right depends on finding the right upper and lower bounds for each class.

Example:

Draw a histogram to represent the data below showing the masses of parcels (given to the nearest 100 g).

Mass of parcel (to nearest 100 g)	100 - 200	300 - 400	500 - 700	800 - 1100
Number of parcels	100	250	600	50

First draw a table showing the upper and lower class bounds, plus the frequency density:

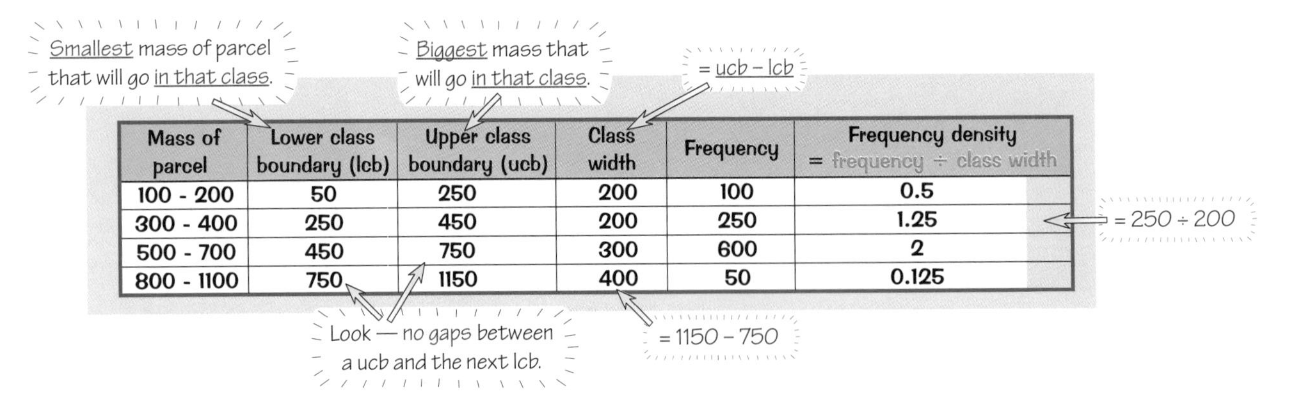

Mass of parcel	Lower class boundary (lcb)	Upper class boundary (ucb)	Class width	Frequency	Frequency density = frequency ÷ class width
100 - 200	50	250	200	100	0.5
300 - 400	250	450	200	250	1.25
500 - 700	450	750	300	600	2
800 - 1100	750	1150	400	50	0.125

Now you can draw the histogram.

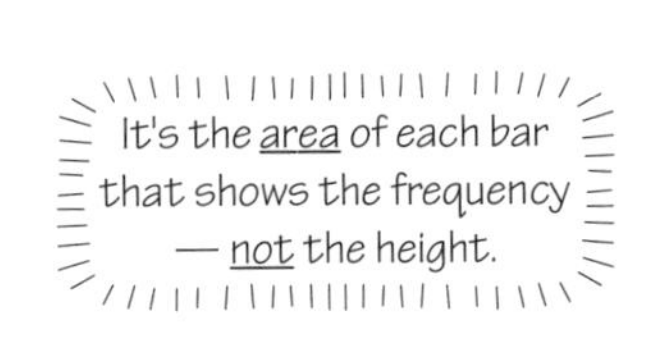

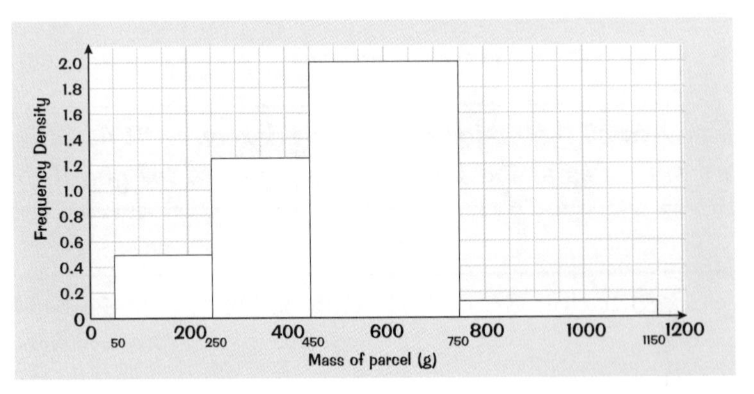

Note: A class with a lower class boundary of 50 g and upper class boundary of 250 g can be written in different ways.

So you might see: "100 - 200 to nearest 100 g"

"50 ≤ mass < 250"

"50–", followed by "250–" for the next class and so on.

They all mean the same — just make sure you know how to spot the lower and upper class boundaries.

Stem and Leaf Diagrams

Stem and Leaf Diagrams look nothing like stems or leaves

They're just an easy way to represent your data. And they come in two flavours — plain or back-to-back.

Example: The lengths in metres of cars in a car park were measured to the nearest 10 cm. Draw a stem and leaf diagram to show the following data: 2.9, 3.5, 4.0, 2.8, 4.1, 3.7, 3.1, 3.6, 3.8, 3.7

It's best to do a rough version first, and then put the 'leaves' in order afterwards.

It's a good idea to cross out the numbers (in pencil) as you add them to your diagram.

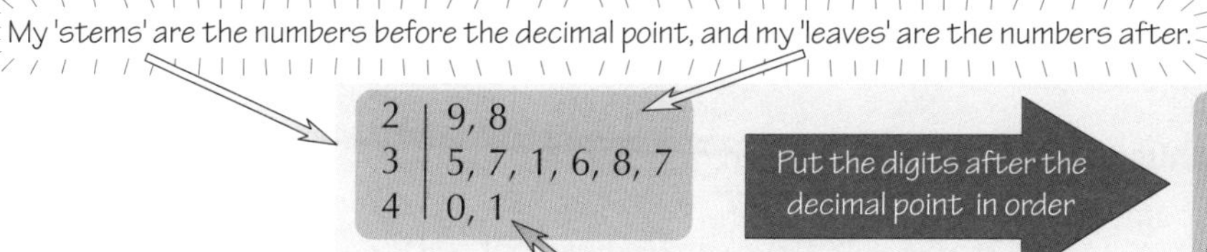

Digits after the decimal point — this row represents 4.0 m and 4.1 m.

Put the digits after the decimal point in order

2 | 8, 9
3 | 1, 5, 6, 7, 7, 8
4 | 0, 1

Key 2|9 means 2.9 m

Always give a key.

Example: The heights of boys and girls in a year 11 class are given to the nearest cm in a back-to-back stem and leaf diagram below. Write out the data in full.

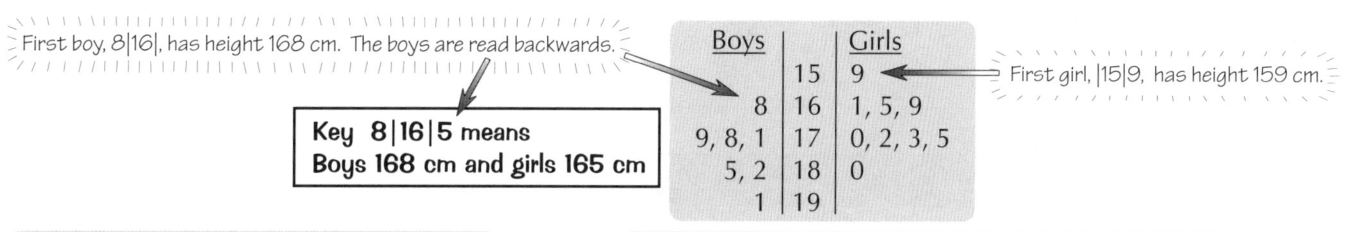

Boys: 168, 171, 178, 179, 182, 185, 191

Girls: 159, 161, 165, 169, 170, 172, 173, 175, 180

Practice Questions

1) *The stem and leaf diagram on the right represents the lengths (in cm) of 15 bananas. Write down the original data as a list.*

12 | 8
13 | 2, 5
14 | 3, 3, 6, 8
15 | 2, 9
16 | 1, 1, 2, 3
17 | 0, 2

Key 12|8 means 12.8 cm

2) *Construct a back-to-back stem and leaf diagram to represent the following data:*
Boys' test marks 34, 27, 15, 39, 20, 26, 32, 37, 19, 22
Girls' test marks 21, 38, 37, 12, 27, 28, 39, 29, 25, 24, 31, 36

3) *Twenty phone calls were made by a householder one evening. The lengths of the calls (in minutes to the nearest minute) are recorded below. Draw a histogram of the data.*

Length of call	0 - 2	3 - 5	6 - 8	9 - 15
Number of calls	10	6	3	1

Sample exam question:

4) The profits of 100 businesses are given in the table.

Profit, £x million.	Number of businesses
$4.5 \leqslant x < 5.0$	24
$5.0 \leqslant x < 5.5$	26
$5.5 \leqslant x < 6.0$	21
$6.0 \leqslant x < 6.5$	19
$6.5 \leqslant x < 8.0$	10

(a) Represent the data in a histogram. [3 marks]

(b) Comment on the distribution of the profits of the businesses. [2 marks]

First things first: remember — there are lies, damned lies and statistics...

Histograms shouldn't really cause too many problems — this is quite a friendly topic really. The main things to remember are to work out the lower and upper boundaries of each class properly, and then make sure you use frequency density (rather than just the frequency). Stem and leaf diagrams — hah, they're easy, I do them in my sleep. Make sure you can too.

Mean, Median, Mode and Range

Skip these two pages if you're doing AQA A S1.

*The **Definitions** are really GCSE stuff*

You more than likely already know them. But if you don't, learn them now — you'll be needing them loads.

Mean = $\bar{x} = \frac{\sum x}{n}$ or $\frac{\sum fx}{n}$

The Σ (sigma) things just mean you add stuff up — so Σx means you add up all the values of x.

where each x is a data value, f is the frequency of each x (the number of times each one occurs), and n is the total number of data values.

Median = middle data value when all the data values are placed in order of size.

This will be the $\left(\frac{n+1}{2}\right)$th value in the ordered list.

Mode = most frequently occurring data value.

Range = highest value – lowest value

Example: Find the mean, median, mode and range of the following list of data: 2, 3, 6, 2, 5, 9, 3, 8, 7, 2

Put in order first: 2, 2, 2, 3, 3, 5, 6, 7, 8, 9

Mean $= \frac{2+2+2+3+3+5+6+7+8+9}{10} = \underline{4.7}$

Median = average of 5th and 6th values = average of 3 and 5 = $\underline{4}$

Since $\frac{n+1}{2} = 5.5$

Mode = $\underline{2}$

Range = 9 – 2 = $\underline{7}$

*Use a **Table** when there are a lot of **Numbers***

Example:

The numbers of letters received one day in 100 houses was recorded. Find the mean, median, mode and range number of letters.

Number of letters	Number of houses
0	11
1	25
2	27
3	21
4	9
5	7

The first thing to do is make a table like this one:

Number of letters x	Number of houses f		fx
0	11	(11)	0
1	25	(36)	25
2	27	(63)	54
3	21		63
4	9		36
5	7		35
totals	100		213

Multiply x by f to get this column.

Put the running total in brackets — it's handy when you're finding the median. (But you can stop when you get past halfway.)

$n = \sum f = 100$

$\sum fx = 213$

1. The mean is easy — just divide the total of the fx-column by the total of the f-column (= n). **Mean** $= \frac{213}{100}$ **= 2.13 letters**

2. To find the position of the median, add 1 to the total frequency (= n) and then divide by 2.
Here the median is in position: (100 + 1) ÷ 2 = $\underline{50.5}$.
So the median is halfway between the 50th and 51st data values.
Using your running total of f, you can see that the data values in positions 37 to 63 are all 2s. This means the data values at positions 50 and 51 are both 2 — so **Median = 2 letters**

3. The highest frequency is for 2 letters — so **Mode = 2 letters**

4. Range = highest data value – lowest data value. So **Range = 5 – 0 = 5 letters**

Mean, Median, Mode and Range

*If the data's **Grouped** you'll have to **Estimate***

If the data's grouped, you can only estimate the mean, median and mode.

There are no precise readings here — each reading's been put into one of these groups.

Example: The height of a number of trees was recorded. The data collected is shown in this table:

Height of tree to nearest m	0 - 5	6 - 10	11 - 15	16 - 20
Number of trees	26	17	11	6

Find an estimate of the mean height of the trees.

Here, you assume that every reading in that class takes the mid-class value (which you find by adding the lower class boundary to the upper class boundary and dividing by 2). It's best to make another table...

Height of tree to nearest m	mid-class value x	Number of trees f	fx
0 - 5	2.75	26 (26)	71.5
6 - 10	8	17 (43)	136
11 - 15	13	11	143
16 - 20	18	6	108
	Totals	60 (= n)	458.5(= Σfx)

Lower class boundary = 0.
Upper class boundary = 5.5.
So the mid-class value = (0 + 5.5) ÷ 2 = 2.75.

Estimated mean $= \frac{458.5}{60} = 7.64$ m

*Estimate the **Median** by assuming the values are **Evenly Spread***

The median position here is (60 + 1) ÷ 2 = 30.5, so the median is the 30.5th reading (halfway between the 30th and 31st). Your 'running total' tells you the median must be in the '6 - 10' class.
Now you have to assume that all the readings in this class are evenly spread.

There are 26 trees before class 6 - 10, so the 30.5th tree is the 4.5th value of this class.
Divide the class into 17 equally wide parts (as there are 17 readings) and assume there's a reading at each new point.
Then you want the reading that's 4.5 parts along.

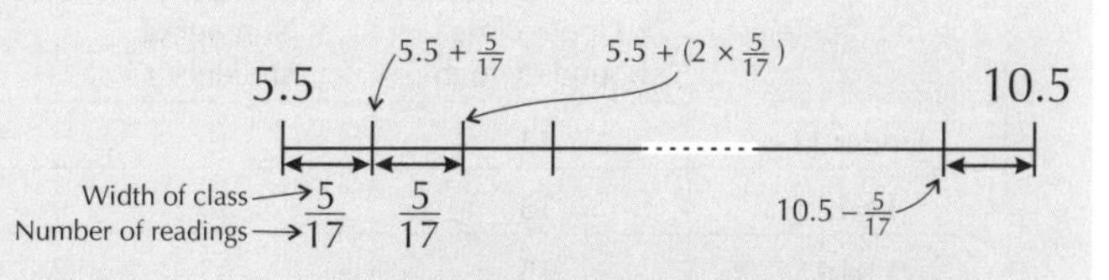

So the median = lower class boundary + (4.5 × width of each 'bit')

$= 5.5 + \left(4.5 \times \frac{5}{17}\right) = 6.8$ m

The modal class is the class with most readings in it. In this example the modal class is 0 - 5 m.

Practice Questions

1) Calculate the mean, median and mode of the data in the table on the right.

x	0	1	2	3	4
f	5	4	4	2	1

2) The speeds of 60 cars travelling in a 40 mph speed limit area were measured to the nearest mph. The data is summarised in the table. Calculate estimates of the mean and median, and state the modal class.

Speed (mph)	30 - 34	35 - 39	40 - 44	45 - 50
Frequency	12	37	9	2

Sample Exam Question:

3) The stem and leaf diagram shows the test marks for 30 male students and 16 female students.

(a) Find the median test mark of the male students. [1 mark]

(b) Compare the distribution of the male and female marks. [2 marks]

Male students		Female students
8, 3, 3	4	
8, 7, 7, 7, 5, 3, 2	5	5, 6, 7
9, 7, 6, 6, 5, 5, 2, 2, 1, 1, 0	6	1, 2, 3, 3, 4, 5, 6, 7, 9
9, 9, 8, 5, 4, 3, 1, 0, 0	7	2, 4, 8, 9

Key 5|6|2 means Male student test mark 65 and Female student test mark 62

I can't deny it — this page really is kind of average...

Doing all this stuff isn't that hard — it's remembering all the different names that gives me a headache. But it's all made easier if you learn that the MeDian is the one in the MiDdle, while the MOde is the one that there's MOst of. The mean, well, that's just your common or garden 'average' that you learnt about while you were still in short trousers.

Cumulative Frequency Diagrams

Skip these two pages if you're doing AQA A S1.

Quartiles divide the data into Four

The median divides the data into two — the quartiles divide the data into four.

Example: Find the median and quartiles of the following data: 2, 5, 3, 11, 6, 7, 1

First put the list in order: 1 2 3 5 6 7 11

There are 7 numbers, so the median is in position 4 (i.e. you take the fourth number along): Median = 5

The middle of the set of numbers below the median is the lower quartile (Q_1) and the middle of the set above the median is the upper quartile (Q_3). ⟹ 1 2 3 5 6 7 11 (Q_1 = 2, Q_3 = 7)

About 25% of the readings are less than the lower quartile. About 75% are less than the upper quartile.

Lower quartile = 2 Upper quartile = 7

Use Cumulative Frequency Graphs to find the Median and Quartiles

Cumulative frequency means 'running total'. Cumulative frequency diagrams make medians and quartiles easy to find...

Example: The ages of 200 students in a school are recorded in the table below.

Draw a cumulative frequency graph and use it to estimate the median age and the interquartile range.

Also estimate how many students are older than 18.

Age in completed years	11 - 12	13 - 14	15 - 16	17 - 18
Number of students	50	65	58	27

(1) First draw a table showing the upper class boundaries and the cumulative frequency:

Age in completed years	Upper class boundary (ucb)	Number of students, f	Cumulative frequency (cf)
Under 11	11	0	0
11-12	13	50	50
13-14	15	65	115
15-16	17	58	173
17-18	19	27	200

The first reading in a cumulative frequency table must be zero — so add this extra row to show the number of students with age less than 11 is 0.

CF is the number of students with age less than the ucb — it's the same thing as your running total from the last two pages.

The last number in the CF column should always be the total number of readings.

People say they're '18' right up until their 19th birthday — so the ucb of class 17-18 is 19.

Next draw the axes — cumulative frequency always goes on the vertical axis. Here, age goes on the other axis. Then plot the upper class boundaries against the cumulative frequencies, and join the points.

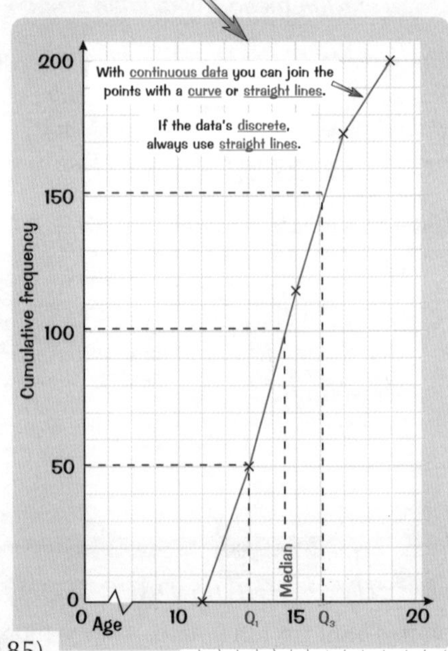

Always plot the upper class boundary of each class.

(2) To find the median from a graph, go to the median position on the vertical scale and read off the value from the horizontal axis.

Median position $= \frac{1}{2}(200+1) = 100.5$ so Median = 14.5 years

Then you can find the **quartiles** in the same way. Find their positions first:

Q_1 position $= \frac{1}{4}\times(200+1) = 50.25$ (i.e. between the 50th and 51st readings)

Q_3 position $= \frac{3}{4}\times(200+1) = 150.75$ (between the 150th and 151st readings)

Lower quartile, Q_1 = 13 years Upper quartile, Q_3 = 16.2 years

The **interquartile range** (IQR) $= Q_3 - Q_1$. It measures spread.
The smaller it is the less spread the data is.

IQR $= Q_3 - Q_1 = 16.2 - 13 = 3.2$ years

(3) To find how many students are older than 18, first go up from 18 on the horizontal axis, and read off the number of students younger than 18 (= 185).

Then the number of students older than 18 is just 200 – 185 = 15 (approximately)

Cumulative Frequency Diagrams

Percentiles divide the data into 100

Percentiles divide the data into 100 — the median is the 50th percentile and Q_1 is the 25th percentile, etc.

Example: The position of the 11th percentile (P_{11}) is $\frac{11}{100}\times$(total frequency +1) = $\frac{11}{100}\times 201 = 22.11$

Or by going $\frac{11}{100}$ up the graph, you can see the 11th percentile is about 12.

You find interpercentile ranges by subtracting two percentiles, e.g. the middle 60% of the readings = $P_{80} - P_{20}$.

Box and Whisker Diagrams are useful for comparing distributions

Box and whisker plots show the median and quartiles in an easy-to-look-at kind of way...
These are sometimes called box plots.

They look like this:

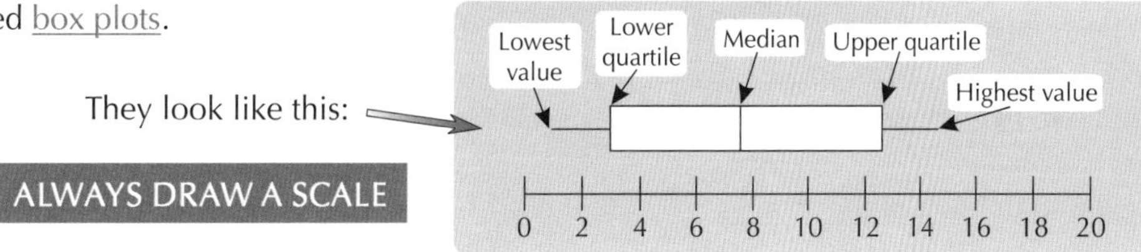

ALWAYS DRAW A SCALE

Practice Questions

1) ***Draw a cumulative frequency diagram of the data given in the table. Use your diagram to estimate the median and interquartile range.***

Distance walked (km)	0 - 2	2 - 4	4 - 6	6 - 8
Number of walkers	10	5	3	2

Sample Exam questions:

2) A shopkeeper records the age, to the nearest year, of the customers that enter his shop before 9:00 am one morning.

Age of customer to nearest year	5 - 10	11 - 15	16 - 20	21 - 30	31 - 40	41 - 70
Number of customers	2	3	10	2	2	1

(a) On graph paper, draw a cumulative frequency diagram. [4 marks]

(b) From your graph estimate
  (i) the median age of the customers. [1 mark]
  (ii) the number of customers at least 12 years old. [2 marks]

3) Two workers iron clothes. Each irons 10 items, and records the time it takes them for each, to the nearest minute.

Worker A: 3 5 2 7 10 4 5 5 4 12
Worker B: 3 4 8 6 7 8 9 10 11 9

(a) For worker A's times. Find:
  (i) the median, [1 mark]
  (ii) the lower and upper quartiles. [2 marks]

(b) On graph paper draw, using the same scale, two box plots to represent the times of each worker. [6 marks]

(c) Make one statement comparing the two sets of data. [1 mark]

(d) Which worker would be best to employ? Give a reason for you answer. [1 mark]

WG Grace — an old-time box and whisker man...

Cumulative frequency sounds a bit scarier than running total — but if you remember that they're the same thing, that'll help. You might get a question on quartiles... to find these, you have to find the median of the bottom half and the median of the top half of the data — not including the middle value in either half if you have an odd number of values.

Variance and Standard Deviation

Skip this page if you're doing AQA A S1.

Standard deviation and variance both measure how spread out the data is from the mean — the bigger the variance, the more spread out your readings are.

The *Formulas* look pretty *Tricky*

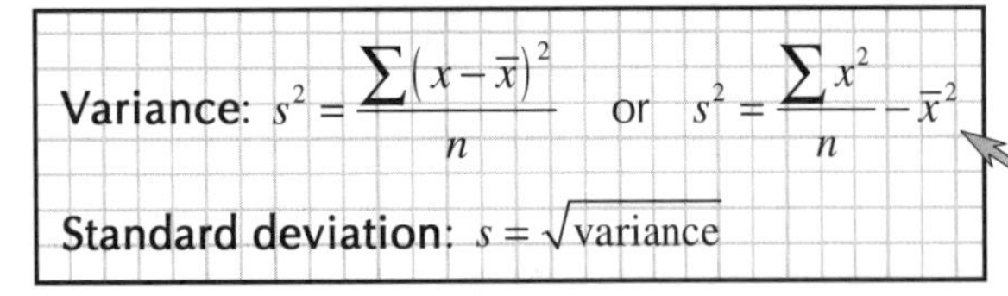

Variance: $s^2 = \frac{\sum(x-\bar{x})^2}{n}$ or $s^2 = \frac{\sum x^2}{n} - \bar{x}^2$

Standard deviation: $s = \sqrt{\text{variance}}$

The x-values are the data, $\bar{x}$ is the mean, and n is the number of data values.

The second formula is easier to use.

Example: Find the mean and standard deviation of the following numbers: 2, 3, 4, 4, 6, 11, 12

1) Find the total of the numbers first: $\sum x = 2+3+4+4+6+11+12 = 42$

2) Then the mean is easy: $\text{Mean} = \bar{x} = \frac{\sum x}{n} = \frac{42}{7} = 6$

3) Next find the sum of the squares: $\sum x^2 = 4+9+16+16+36+121+144 = 346$

4) Use this to find the variance: Variance, $s^2 = \frac{\sum x^2}{n} - \bar{x}^2 = \frac{346}{7} - 6^2 = 49.43 - 36 = 13.43$

5) And take the square root to find the standard deviation: Standard deviation $= \sqrt{13.43} = 3.66$ to 3 sig. fig.

Questions about Standard Deviation *can look a bit* Weird

They can ask questions about standard deviation in different ways. But you just need to use the same old formulas.

Example:

The mean of 10 boys' heights is 180 cm, and the standard deviation is 10 cm. The mean for 9 girls is 165 cm, and the standard deviation is 8 cm. Find the mean and standard deviation of the whole group of 19 girls and boys.

① Let the boys' heights be x and the girls' heights be y.

Write down the formula for the mean and put the numbers in for the boys: $\bar{x} = \frac{\sum x}{n} \Rightarrow 180 = \frac{\sum x}{10} \Rightarrow \sum x = 1800$

Do the same for the girls: $165 = \frac{\sum y}{9} \Rightarrow \sum y = 1485$

So the sum of the heights for the boys and the girls $= \sum x + \sum y = 1800 + 1485 = 3285$

And the mean height of the boys and the girls is: $\frac{3285}{19} = 172.9 \text{ cm}$

Round the fraction to 1 dp to give your answer. But if you need to use the mean in more calculations, use the fraction (or your calculator's memory) so you don't lose accuracy.

② Now for the variance — write down the formula for the boys first: $s_x^2 = \frac{\sum x^2}{n} - \bar{x}^2 \Rightarrow 10^2 = \frac{\sum x^2}{10} - 180^2 \Rightarrow \sum x^2 = 10\times(100+32400) = 325000$

Do the same for the girls: $s_y^2 = \frac{\sum y^2}{n} - \bar{y}^2 \Rightarrow 8^2 = \frac{\sum y^2}{9} - 165^2 \Rightarrow \sum y^2 = 9\times(64+27225) = 245601$

Okay, so the sum of the squares of the heights of the boys and the girls is: $\sum x^2 + \sum y^2 = 325000 + 245601 = 570601$

Which means the variance of all the heights is: $s^2 = \frac{570601}{19} - \left(\frac{3285}{19}\right)^2 = 139.0 \text{ cm}^2$

Don't use the rounded mean (172.9) — you'll lose accuracy.

And finally the standard deviation of the boys and the girls is: $s = \sqrt{139.0} = 11.8 \text{ cm}$

Phew.

Variance and Standard Deviation

Skip 'mid-class values' if you're doing AQA B S1.

Use Mid-Class Values if your data's in a Table

With grouped data, assume every reading takes the mid-class value. Then use the frequencies to find $\sum x$ and $\sum x^2$.

Example: The heights of sunflowers in a garden were measured and recorded in the table below. Estimate the mean height and the standard deviation.

Height of sunflower	$150 \le x < 170$	$170 \le x < 190$	$190 \le x < 210$	$210 \le x < 230$
Number of sunflowers	5	10	12	3

Draw up another table, and include columns for the mid-class values x, as well as fx and fx^2:

Height of sunflower	Mid-class value, x	x^2	f	fx	fx^2
$150 \le x < 170$	160	25600	5	800	128000
$170 \le x < 190$	180	32400	10	1800	324000
$190 \le x < 210$	200	40000	12	2400	480000
$210 \le x < 230$	220	48400	3	660	145200
		Totals	30 (= n)	5660 (= Σx)	1077200 (= Σx^2)

fx^2 means $f \times (x^2)$ — not $(fx)^2$.

Now you've got the totals in the table, you can calculate the mean and variance:

$$\text{Mean} = \bar{x} = \frac{\sum x}{n} = \frac{5660}{30} = 189 \text{ to 3 sig. fig.}$$

$$\text{Variance} = s^2 = \frac{\sum x^2}{n} - \bar{x}^2 = \frac{1077200}{30} - \bar{x}^2 = 312 \text{ to 3 sig. fig.}$$

$$\text{Standard deviation} = \sqrt{312} = 17.7 \text{ to 3 sig. fig.}$$

Practice Questions

1) **Find the mean and standard deviation of the following numbers: 11, 12, 14, 17, 21, 23, 27.**

2) **The scores in an IQ test for 50 people are recorded in the table below.**

Score	100 - 106	107 - 113	114 - 120	121 - 127	128 - 134
Frequency	6	11	22	9	2

Calculate the mean and variance of the distribution.

Sample Exam question:

3) In a supermarket two types of chocolate drops were compared.
The weights (in grams) of 20 chocolate drops of brand A are summarised by:

$$\sum A = 60.3 \text{ g} \qquad \sum A^2 = 219 \text{ g}^2$$

The mean weight of 30 chocolate drops of brand B was 2.95 g, and the standard deviation was 1 g.

(a) Find the mean weight of a brand A chocolate drop. [1 mark]

(b) Find the standard deviation of the weight of the brand A chocolate drops. [3 marks]

(c) Compare brands A and B. [2 marks]

(d) Find the standard deviation of the weight of all 50 chocolate drops. [4 marks]

People who enjoy this stuff are standard deviants...

The formula for the variance looks pretty scary, what with the S's and $\bar{x}$'s floating about. But it comes down to 'the mean of the squares minus the square of the mean'. That's how I remember it anyway — and my memory's rubbish.
Ooh, while I remember... don't forget to work out mid-class values carefully, using the upper and lower class boundaries.

Coding

Skip these two pages if you're doing OCR A S1 or AQA A S1.

Coding means doing something to every reading (like adding a number, or multiplying by a number) to make life easier.

Coding can make the Numbers much Easier

Finding the mean of 1001, 1002 and 1006 looks hard(ish). But take 1000 off each number and finding the mean of what's left (1, 2 and 6) is much easier — it's 3. So the mean of the original numbers must be 1003. That's coding.

You usually change your original variable x to an easier one to work with y (so here, if $x = 1001$, then $y = 1$.)

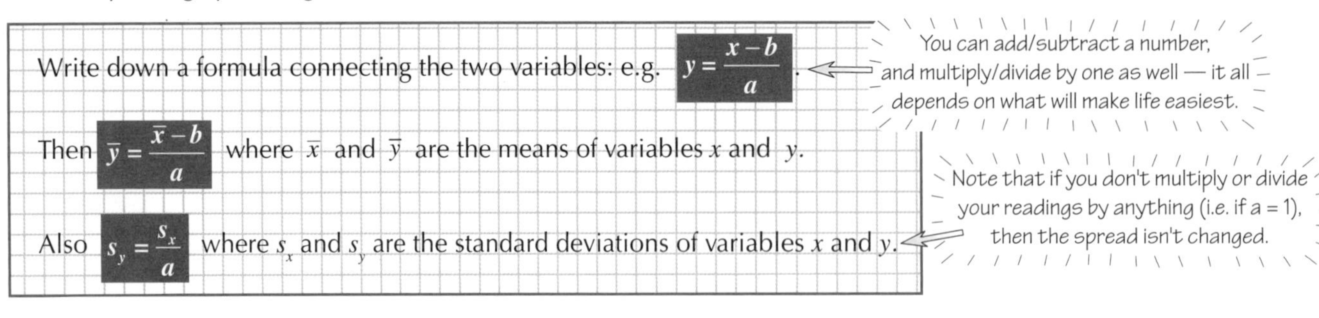

Write down a formula connecting the two variables: e.g. $y = \frac{x-b}{a}$

Then $\bar{y} = \frac{\bar{x}-b}{a}$ where $\bar{x}$ and $\bar{y}$ are the means of variables x and y.

Also $s_y = \frac{s_x}{a}$ where s_x and s_y are the standard deviations of variables x and y.

You can add/subtract a number, and multiply/divide by one as well — it all depends on what will make life easiest.

Note that if you don't multiply or divide your readings by anything (i.e. if a = 1), then the spread isn't changed.

Example: Find the mean and standard deviation of: 1 000 020, 1 000 040, 1 000 010 and 1 000 050.

The obvious thing to do is subtract a million from every reading to leave 20, 40, 10 and 50.
Then make life even simpler by dividing by 10 — giving 2, 4, 1 and 5.

1. So use the coding: $y = \frac{x-1000000}{10}$. Then $\bar{y} = \frac{\bar{x}-1000000}{10}$ and $s_y = \frac{s_x}{10}$.

2. Find the mean and standard deviation of the y values: $\bar{y} = \frac{2+4+1+5}{4} = 3$ $\quad s_y = \frac{2^2+4^2+1^1+5^2}{4} - 3^2 = \frac{46}{4} - 9 = 2.5$

3. Then use the formulas to find the mean and standard deviation of the original values:

   $\bar{x} = 10\bar{y} + 1000000 = (10\times3) + 1000000 = 1000030$ $\quad s_x = 10s_y = 10\times2.5 = 25$

You can use coding with Summarised Data

This kind of question looks tricky at first — but use the same old formulas and it's a piece of cake.

Example: A set of 10 numbers, x, can be summarised as shown: $\sum(x-10) = 15$ and $\sum(x-10)^2 = 100$
Find the mean and standard deviation of x.

1. Okay, the obvious first thing to try is: $y = x - 10$

   That means: $\sum y = 15$ and $\sum y^2 = 100$

2. Work out $\bar{y}$ and s_y^2 using the normal formulas: $\bar{y} = \frac{\sum y}{n} = \frac{15}{10} = 1.5$

   $s_y^2 = \frac{\sum y^2}{10} - \bar{y}^2 = \frac{100}{10} - 1.5^2 = 10 - 2.25 = 7.75$

   so $s_y = 2.78$ to 3 sig. fig.

3. Then finding the mean and standard deviation of the x-values is easy: $\bar{x} = \bar{y} + 10 = 1.5 + 10 = 11.5$

   $s_x = s_y = 2.78$ to 3 sig. fig.

   The spread of x is the same as the spread of y since you've only subtracted 10 from every number.

Coding

Sensible Coding can make life Much Easier

Find the mean and standard deviation of the data in this table:

Class	10 - 19	20 - 29	30 - 39
f	2	5	3

It's grouped data, so use mid-class values — these are $x = 14.5$, 24.5 and 34.5.

Now let $y = \frac{x - 24.5}{10}$ and then draw up another table: ⇐ This coding will make all the numbers in the table dead easy.

Class	Mid-class x	$y = \frac{x-24.5}{10}$	f	fy	fy^2
10 - 19	14.5	-1	2	-2	2
20 - 29	24.5	0	5	0	0
30 - 39	34.5	1	3	3	3
		Totals	10 (= n)	1 (= Σy)	5 (= Σy^2)

$\bar{y} = \frac{\sum y}{n} = \frac{1}{10} = 0.1$ and $s_y^2 = \frac{\sum y^2}{10} - \bar{y}^2 = \frac{5}{10} - 0.1^2 = 0.5 - 0.01 = 0.49$

So $s_y = 0.7$

But $y = \frac{x - 24.5}{10}$ so $\bar{y} = \frac{\bar{x} - 24.5}{10}$, which means $\bar{x} = 10\bar{y} + 24.5 = (10 \times 0.1) + 24.5 = 25.5$

Since everything has been divided by 10, the spread of y is not the same as the spread of x.

In fact, $s_y = \frac{s_x}{10}$ so $s_x = 10 s_y = 7$

Practice Questions

1) ***For a set of data,*** $n = 100$, $\sum(x - 20) = 125$, ***and*** $\sum(x - 20)^2 = 221$.
Find the mean and standard deviation of x.

2) ***The time taken (to the nearest minute) for a commuter to travel to work on 20 consecutive days is recorded in the table. Use coding to find the mean and standard deviation of the times.***

Time to nearest minute	30 - 33	34 - 37	38 - 41	42 - 45
Frequency	3	6	7	4

Sample Exam question:

3) A group of 19 people played a game. The scores, x, that the people achieved are summarised by:

$$\sum(x - 30) = 228 \text{ and } \sum(x - 30)^2 = 3040$$

(a) Calculate the mean and the standard deviation of the 19 scores. [3 marks]

(b) Show that $\sum x = 798$ and $\sum x^2 = 33820$. [3 marks]

(c) Another student played the game. Her score was 32.
Find the new mean and standard deviation of all 20 scores. [4 marks]

I thought the coding page would be a little more... well, James Bond...

Coding data isn't hard — the only tricky thing can be to work out how best to code it, although there will usually be some pretty hefty clues in the question if you care to look. But remember that adding/subtracting a number from every reading won't change the spread (the variance or standard deviation), but multiplying/dividing readings by something will.

Skewness and Outliers

Skip these two pages if you're doing OCR A S1 or AQA A S1. Skip 'skewness' (but not 'outliers') if you're doing AQA B S1.

Skewness tells you whether your data is symmetrical — or kind of lopsided.

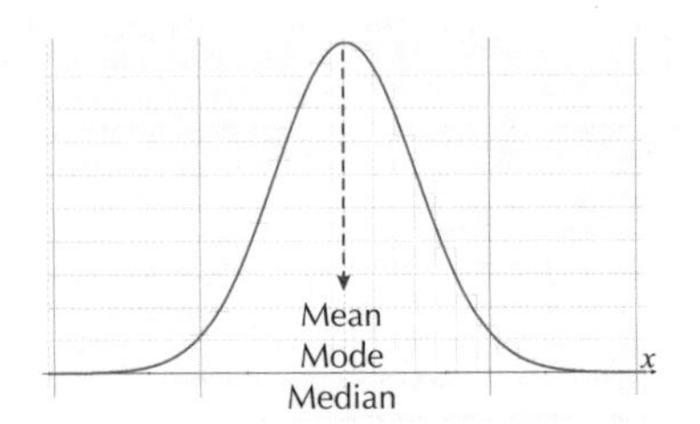

This is a typical symmetrical distribution.

Notice: mean = median = mode

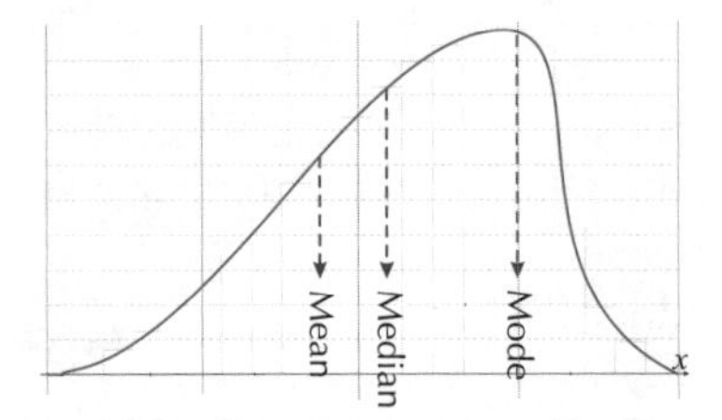

A negatively skewed distribution has a tail on the left. Most data values are on the higher side.

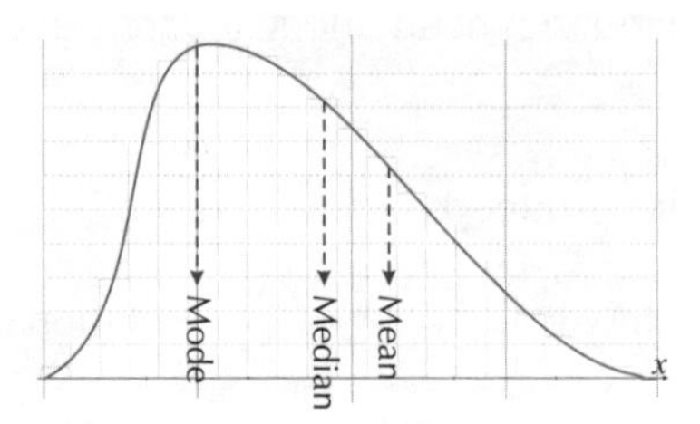

A positively skewed distribution has a tail on the right. Most data values are on the lower side.

For all distributions: **mean – mode = 3 × (mean – median)** — approximately.

Measure skewness using *Pearson's Coefficient of Skewness...*

A coefficient of skewness measures how 'all-up-one-end' your data is. You need to know a couple of formulas...

$$\text{Pearson's coefficient of skewness} = \frac{\text{mean} - \text{mode}}{\text{standard deviation}} = \frac{3(\text{mean} - \text{median})}{\text{standard deviation}}$$

This usually lies between -3 and +3.
So if Pearson's coefficient of skewness is -0.1, then the distribution is slightly negatively skewed.

...or the *Quartile Coefficient of Skewness*

Remember that Q_1 is the lower quartile, Q_3 is the upper quartile, and the median is Q_2.

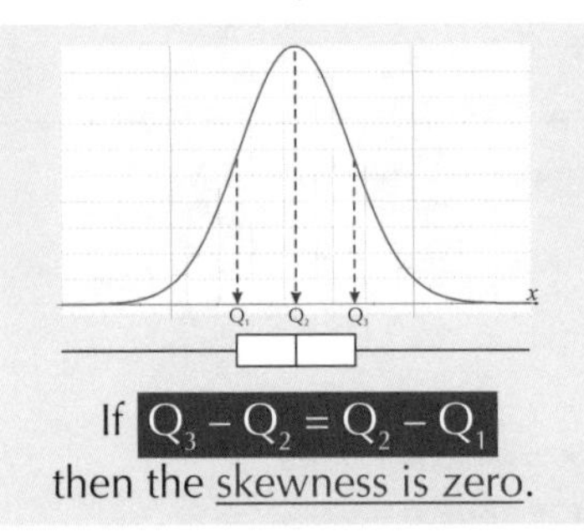

If $Q_3 - Q_2 = Q_2 - Q_1$ then the skewness is zero.

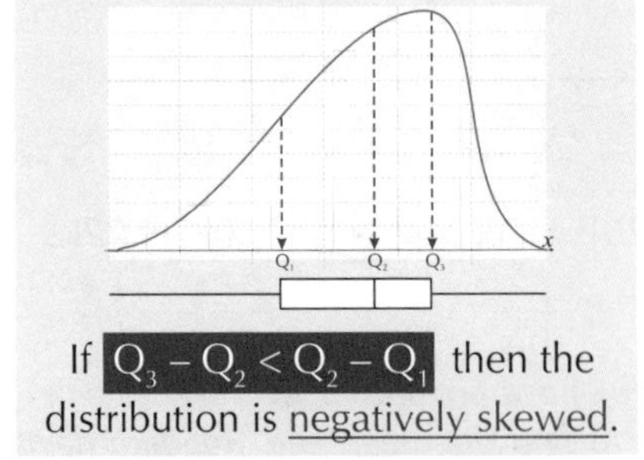

If $Q_3 - Q_2 < Q_2 - Q_1$ then the distribution is negatively skewed.

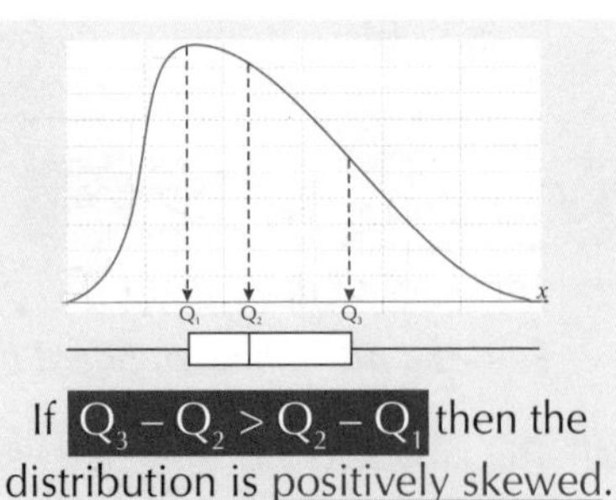

If $Q_3 - Q_2 > Q_2 - Q_1$ then the distribution is positively skewed.

$$\text{Quartile coefficient of skewness} = \frac{(Q_3 - Q_2) - (Q_2 - Q_1)}{Q_3 - Q_1} = \frac{Q_3 - 2Q_2 + Q_1}{Q_3 - Q_1}$$

Example: This table summarises the marks obtained in Maths 'calculator' and 'non-calculator' papers.

Calculate the Pearson's and Quartile coefficient of skewness for each paper. Comment on the distributions.

Calculator Paper		Non-calculator paper
40	Lower quartile, Q_1	35
58	Median, Q_2	42
70	Upper quartile, Q_3	56
55	Mean	46.1
21.2	Standard deviation	17.8

The quartile coefficient of skewness tells you that the calculator paper scores are slightly negatively skewed and that the non-calculator paper scores are positively skewed.

Calculator Paper		Non-calculator Paper
$\frac{3\times(55.0-58)}{21.2} = \frac{-9}{21.2} = -0.425$	Pearson's coefficient of skewness	$\frac{3\times(46.1-42)}{17.8} = \frac{12.3}{17.8} = 0.691$
$\frac{70-2\times58+40}{70-40} = \frac{-6}{30} = -0.2$	Quartile coefficient of skewness	$\frac{56-2\times42+35}{56-35} = \frac{7}{21} = 0.333$

Generally students have done better (compared to the mean) on the calculator paper.
Pearson's coefficient of skewness confirms these results.

Skewness and Outliers

An outlier is a freak piece of data that lies a long way from the rest of the readings.
To find whether a reading is an outlier you have to measure how far away from the rest of the data it is.

Outliers fall Outside Fences

There are various ways to decide if a reading is an outlier — the method you should use is always described in the question.

Example: A data value is considered to be an outlier if it is more than 3 times the IQR above the upper quartile or more than 3 times the IQR below the lower quartile.

The lower and upper quartiles of a data set are 70 and 100. Decide whether the data values 20 and 210 are outliers.

First you need the IQR: $Q_3 - Q_1 = 100 - 70 = 30$

Then it's a piece of cake to find where your fences are.

Lower fence first: $Q_1 - (3 \times \text{IQR}) = 70 - (3 \times 30) = -20$

And the upper fence: $Q_3 + (3 \times \text{IQR}) = 100 + (3 \times 30) = 190$

-20 and 190 are called fences. Any reading lying outside the fences is considered an outlier.

20 is inside the lower fence, so it is not an outlier. 210 is outside the upper fence, so it is an outlier.

Practice Questions

1) ***A data value is considered an outlier if it's more than 3 times the IQR above the upper quartile or more than 3 times the IQR below the lower quartile.***
If the lower and upper quartiles of a data set are 62 and 88, decide which of the following data are outliers:
a) 161, b) 176, c) 0

2) ***Find the median and quartiles of the data below. Draw a box and whisker diagram, and comment on any skewness.***
Amount of pocket money (in £) received per week by twenty 15-year-olds:
10, 5, 20, 50, 5, 1, 6, 5, 15, 20, 5, 7, 5, 10, 12, 4, 8, 6, 7, 30.

3) ***A set of data has a mean of 10.3, a mode of 10 and a standard deviation of 1.5.***
Calculate Pearson's coefficient of skewness, and draw a possible sketch of the distribution.

Practice Exam Questions:

4) The table shows the number of hits received at a paint ball party.

Age	12	13	14	15	16	17	18	19	20	21	22	23	24	25
Frequency	2	4	6	7	6	4	4	2	1	1	0	0	0	1

(a) Find the median and mode number of hits. [3 marks]
The mean is 16.4 hits.

(b) An outlier is a data value which is greater than $3(Q_3 - Q_1)$ above Q_3 or below Q_1.
Is 25 an outlier? Show your working. [2 marks]

(c) Sketch the shape of the distribution and comment on any skewness. [2 marks]

(d) How would the shape of the distribution be affected if the value of 25 was removed? [1 mark]

5) The data in the table shows the number of mm of rain that fell on 30 days on a tropical island.

mm of rain	5 - 10	10 - 15	15 - 20	20 - 25	25 - 30	30 - 35
No. of days	2	3	5	7	10	3

(a) Draw a cumulative frequency diagram of the data. [3 marks]

(b) Using your diagram estimate the median and quartiles. [3 marks]

(c) Calculate the quartile coefficient of skewness and describe the shape. [2 marks]

'Outlier' is the name I give to something that my theory can't explain...

Those definitions of positive and negative skew aren't the most obvious in the world — and it's easy to get them mixed up. Remember that negative skew involves a tail on the left, which means that a lot of your readings are on the high side. Positive skew is the opposite — a tail on the right, and a bunch of readings that are a little on the low side.

Populations and Sampling

Skip this page if you're doing Edexcel S1 or OCR A S1.

This is about how to estimate things about a group of people or items by just looking at a few of them. It's a pretty wordy topic for maths. But at least it means there are no formulas to get your head round.

In a Census you ask Everyone — in a Survey you ask a Sample of people

The group of people or items you wish to know about is called the population.

This could be: all the students at a university,
all the trees in a forest,
all the bags of sugar produced by a particular company.

If you want to find out about a population, then you could question every person or examine every item. This is called a census. In a census you check everybody or everything.

However, usually you question or examine a sample of the population — this is what you call a survey.

Sampling is the process of picking the people or things to examine — this can be either random or non-random.

Non-random sampling methods include: systematic sampling, quota sampling and cluster sampling.
Random sampling methods include: simple random sampling and stratified random sampling.

Systematic Sampling means choosing items at Regular Intervals

Systematic sampling is a very straightforward method of non-random sampling.

First you list the population in some order.

Next you choose the first member of your sample at random.

Then you choose the rest of your sample by picking every nth item (e.g. you could choose every 10th car or every 5th person on the electoral roll).

Example: Explain how to obtain a systematic sample of size 30 from a population of 1000.

1) Divide your population size by the size of sample you need: $1000 \div 30 = 33\frac{1}{3}$
2) Choose a random start number between 1 and 33, e.g. 11— and include that item in your sample.
3) Find the other members of your sample by adding $33\frac{1}{3}$ each time:

$11 + 33\frac{1}{3} = 44\frac{1}{3}$ — so choose item 44
$44\frac{1}{3} + 33\frac{1}{3} = 77\frac{2}{3}$ — so choose item 78
$77\frac{2}{3} + 33\frac{1}{3} = 111$ — so choose item 111 (and so on until you have 30 items).

Advantages Systematic sampling is good because it's quick and easy to use. Large-scale sampling is often done this way.

Disadvantages But it can give misleading results if there's a periodic cycle in the list. For example, every 7th item on a production line may be faulty — so examining every 7th item wouldn't give an accurate picture.

Quota Sampling is often used in Market Research

Quota sampling is also non-random, and is often used in market research. The population is divided into groups (e.g. age groups, gender groups, income level groups, etc.), and an interviewer is asked to interview a fixed number of people within each group.

The interviewer then stops and asks people (often in a busy shopping area) until they have fulfilled their quota. This is an example of 'opportunity sampling' — only people that want to be interviewed can be questioned.

Advantages The advantage of this method is that it is quick and easy to use.

Disadvantages Quota sampling is completely non-random and the results may not represent the population. For example, all the people who don't want to be interviewed might have something in common.

And where and when the interviewer finds the sample may exclude sections of the population (e.g. looking for a sample in a shopping centre at 10 o'clock on a Wednesday morning will exclude many employed people).

Populations and Sampling

Skip 'cluster sampling' if you're doing Edexcel S1, OCR A S1, AQA A S1 or AQA B S1.

Use **Cluster Sampling** when the population falls naturally into **Subgroups**

1) Cluster sampling is used when the population naturally falls into certain groups — e.g. the population of the UK could be divided into clusters, where each cluster is the population of an individual county.
2) With cluster sampling, it's important that the subgroups or clusters are as similar to each other as possible.

Example: The population might be all the children in secondary education in a particular county.
In this case, the clusters could be the secondary schools.
Then to analyse the children in the population, you could take a random sample of the clusters (i.e. the schools) and obtain information about the children in these schools.

Advantages One advantage of cluster sampling is that it's easier — you only need a list of the children in a few schools, rather than all those in the whole county.
Another is that it is quicker and cheaper — instead of having to go to all the various schools across the county, you'd only have to visit a few.

Disadvantages The disadvantage is that it is not random. It's likely that the children in a particular school will be similar to each other in some way (and this might result in a biased sample).
For example, all the students in a particular school might live in similar kinds of houses, their parents might do similar jobs, etc.

Practice Questions

1) Explain how to get a systematic sample of size 20 from a population of 100.

2) A school held a party for 500 students at the end of term. The school wished to assess the success of the party and decided to obtain information from those who attended the party.
- ***a) Describe the population.***
- ***b) Explain what conducting a census would mean.***
- ***c) Describe one advantage and one disadvantage of doing a census as opposed to a sample survey.***

3) A soap powder manufacturer wishes to find out how many people have watched their television advertisement by conducting interviews in the street. Describe how the method of quota sampling might be used for the interviews.

Sample Exam Question:

4) Listed below are the heights, to the nearest cm, of 25 plants.

20	15	12	8	12
19	20	17	7	12
18	13	14	9	12
13	14	12	16	20
9	10	13	15	17

(a) Calculate the mean of the population. [1 mark]

(b) Take a systematic sample of size 5, clearly explaining your method. [2 marks]

Useful quotes: Get your facts first, then you can distort them as much as you please*...

I try to be honest when I'm writing these little blurby things, and so I'll come straight out with it... this stuff is a bit boring. But — and this is the important point — knowing this stuff is worth marks in the Exam. And easy marks too. Which means that you can afford a little inward smile when you've learnt this stuff well enough to spew it all out onto your Exam paper come the big day. The only slightly tricky bit is cluster sampling — don't confuse it with stratified sampling on page 17.

* Mark Twain

Sampling and Bias

Skip these two pages if you're doing Edexcel S1, OCR A S1 or AQA B S1.

In a random sample each item from the population must have an equal chance of being selected.

*In a **Random Sample** everything has an **Equal Chance** of being picked*

1) With simple random sampling, every person or thing in a population has an equal chance of being in the sample.
2) To get a truly random sample, you need a complete list of the population, and it must include every last person or thing — this isn't always easy to get.
3) To choose the sample, give every member of the population a number and then use a calculator or random number tables to pick the ones to include in your sample.

Example: Use the random number table below to select a sample of 3 people from a population of 80.

8330	3992	1840	0330	1290	3237	9165	4815	0766
2508	9927	6948	8532	1646	1931	8502	8636	2296
9310	0572	1826	3667	6848	3169	6858	9349	4586

1) Give each person in the population a two-digit number from 01 to 80.
2) Roll a dice (or choose some other method) to find a place to start in the random number tables. So if you roll a three, start at the 3rd digit.
3) The first two-digit number is 30, so include item 30 in your sample. The next is 39, so include item 39, and so on.
4) The next two-digit number is 92, but this is no good because it's bigger than 80. So forget 92 and use the next number instead, which is 18.

So the sample of three would be the 30th, 39th and 18th items.

The Ran# button on your calculator can be used in a similar way. Read the instructions to find out more.

*You can sample **With Replacement** or **Without Replacement***

SAMPLING WITH REPLACEMENT is when an item may be chosen for a sample more than once.
SAMPLING WITHOUT REPLACEMENT is when an item may not be chosen more than once.

Example: Find a sample of size 6 from the population of 80 above: a) with replacement, b) without replacement.

Starting at the 3rd digit, the first few numbers would be: 30, 39, 92, 18, 40, 03, 30, 12, 90, 32...
You have to get rid of 92 and 90, making your sample list: 30, 39, 18, 40, 03, 30, 12, 32...

a) If you're sampling with replacement then the sample would be: 30, 39, 18, 40, 03, 30
So choose the items with these numbers. Note that the 30th item has been chosen twice.

b) If you're sampling without replacement then you can't use the 30th item twice.
You'd have to get rid of the second 30. The final sample would be: 30, 39, 18, 40, 03, 12

Sampling without replacement leads to results that are more representative of the population as a whole, i.e. sampling without replacement is more precise than sampling with replacement.

*You can **Reallocate** numbers rather than **Getting Rid** of them*

This method avoids having to get rid of numbers greater than the size of your population (like in the example above, when we got rid of 90 and 92, since the population only had 80 things in it).

Example: Choose a sample, without replacement, from a population of 100.

1) This time you could give each member of the population a three-digit number — from 001 to 100.
2) Instead of disregarding numbers greater than 100, you could use the following rules:
001-100 leave as they are,
101-200 subtract 100,
210-300 subtract 200 etc.,
3) So 239 would become 039, and 184 would become 084, etc.

Sampling and Bias

Stratified Sampling is used when the population has Distinct Groups

Stratified sampling is used when the population is made up of very different groups. It's used to get a sample that has the correct proportions of people or things from all the various groups — but there's still some randomness to it.

Example: An office has 20 female workers and 30 male workers.
Find a sample of size 10 to represent the workers on a committee.

There are two distinct groups — male and female workers. Completely random sampling might lead to a committee of 10 men or 10 women, so stratified sampling is used to get the right proportions of each.

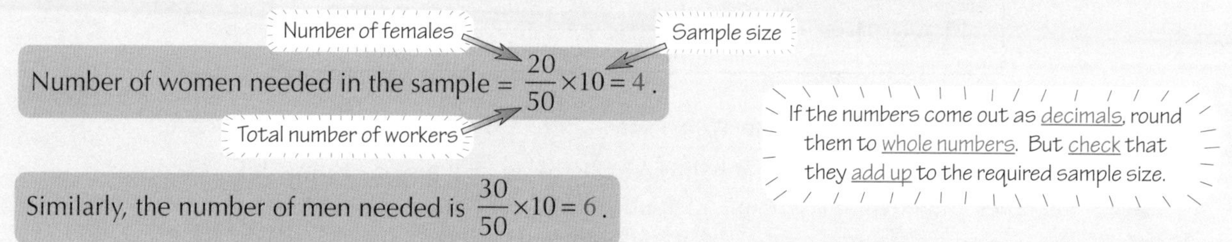

Now you can use a method of random sampling within each group to find the actual people to go on the committee.

Bias can lead to Incorrect Conclusions

Your sample should represent the whole population. If it's biased somehow, then your results could be just plain wrong.

1. Bias can result from picking a sample from an incomplete list of the population.
   For example, the people listed in the phone book will not be a complete list of people living in a certain area.
2. Bias can also result from asking leading questions (questions that 'suggest' the desired answer). An example of a leading question might be, "Do you agree that anyone who likes Product A more than Product B is stupid?"

Practice Questions

1) A sample of 100 items from a population of 1000 items is needed.
Explain how you would obtain a simple random sample without replacement.

2) The following is a list of random numbers: 25808399261.
Use the list to pick the winning raffle ticket when there have been 1000 tickets sold, numbered 1-1000.
Show your method clearly.

Sample Exam question:

3) 25 people in an office were asked how often they shopped at an out-of-town supermarket.

	Number of times the supermarket was visited per month														
Car owners	10	12	4	5	4	2	1	8	9	10	4	4	2	1	4
People without a car	1	0	1	0	2	0	1	0	0	1					

(a) Calculate the mean. [1 mark]
(b) Take a simple random sample, without replacement, of size 10. Show your method clearly. [2 marks]
(c) Explain why this method may not be representative of the population. [1 mark]
(d) An alternative method would be to take a stratified sample of size 10.
  (i) Calculate the numbers of car owners and people without a car that should be included. [2 marks]
  (ii) Take a random stratified sample. Show your methods clearly. [2 marks]
(e) Find both your sample means and make a comparison. [3 marks]

Statistically — surely I should have taken part in an opinion poll by now...

Once again, these pages aren't very interesting, but they aren't very hard either — so be grateful for small mercies. Stratified sampling is easy to confuse with cluster sampling (see p15), but they're different. With cluster sampling, the idea is that the subgroups are as similar to each other as possible — but with stratified sampling, the groups can be very different.

Random Events and Their Probability

If you're doing AQA A S1, this stuff is in the Methods module.

Random events happen by chance. Probability is a measure of how likely they are. It can be a chancy business.

A Random Event has **Various Outcomes**

1) In a trial (or experiment) the things that can happen are called outcomes (so if I time how long it takes to eat my dinner, 63 seconds is a possible outcome).
2) Events are 'groups' of one or more outcomes (so an event might be 'it takes me less than a minute to eat my dinner every day one week').
3) When all outcomes are equally likely, you can work out the probability of an event by counting the outcomes.

$$\text{P(event)} = \frac{\text{Number of outcomes where event happens}}{\text{Total number of possible outcomes}}$$

Example: Suppose I've got a bag with 15 balls in — 5 red, 6 blue and 4 yellow.

If I take a ball out without looking, then any ball is equally likely — there are 15 possible outcomes. Of these 15 outcomes, 5 are red, 6 are blue and 4 are yellow. And so...

$\text{P(red ball)} = \frac{5}{15} = \frac{1}{3}$ $\text{P(blue ball)} = \frac{6}{15} = \frac{2}{5}$ $\text{P(yellow ball)} = \frac{4}{15}$

You can find the probability of either a red or a yellow ball in a similar way...

$\text{P(red or yellow ball)} = \frac{9}{15} = \frac{3}{5}$

The **Sample Space** is the Set of **All Possible Outcomes**

Drawing the sample space (called S) helps you count the outcomes you're interested in.

Example: The classic probability machine is a dice. If you roll it twice, you can record all the possible outcomes in a 6 × 6 table (a possible diagram of the sample space).

There are 36 outcomes in total. You can find probabilities by counting the ones you're interested in (and using the above formula). For example:

(i) The probability of an odd number and then a '1'. There are 3 outcomes that make up this event, so the probability is: $\frac{3}{36} = \frac{1}{12}$

(ii) The probability of the total being 7. There are 6 outcomes that correspond to this event, giving a probability of: $\frac{6}{36} = \frac{1}{6}$

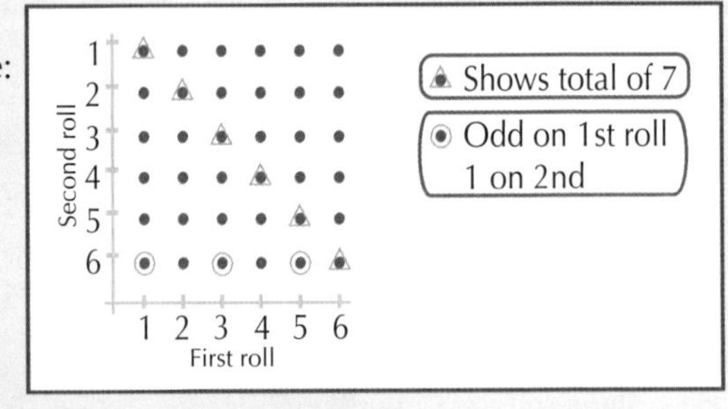

Venn Diagrams show which **Outcomes** correspond to which **Events**

Say you've got 2 events, A and B — a Venn diagram shows which outcomes satisfy event A, which satisfy B, which satisfy both, and which satisfy neither.

(i) All outcomes satisfying event A go in one part of the diagram, and all outcomes satisfying event B go in another bit.
(ii) If they satisfy 'both A and B', they go in the dark green middle bit, written $A \cap B$ (and called the intersection of A and B).
(iii) The whole of the green area is written $A \cup B$ — it means 'either A or B' (and is called the union of A and B).

Again, you can work out probabilities of events by counting outcomes and using the formula above.
You can also get a nice formula linking $P(A \cap B)$ and $P(A \cup B)$.

$$P(A \cup B) = P(A) + P(B) - P(A \cap B)$$

If you just add up the outcomes in A and B, you end up counting $A \cap B$ twice — that's why you have to subtract it.

Example: If you roll a dice, event A could be 'I get an even number', and B 'I get a number bigger than 4'. The Venn diagram would be:

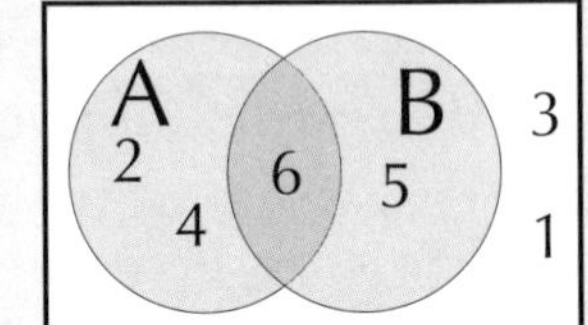

$P(A) = \frac{3}{6} = \frac{1}{2}$ $P(B) = \frac{2}{6} = \frac{1}{3}$ $P(A \cap B) = \frac{1}{6}$ $P(A \cup B) = \frac{4}{6} = \frac{2}{3}$

Here, I've just counted outcomes — but I could have used the formula.

Random Events and Their Probability

Venn Diagrams make it easy to get your head round Tricky Things

Example: A survey was carried out to find what pets people like.

The probability they like dogs is 0.6. The probability they like cats is 0.5. The probability they like gerbils is 0.4. The probability they like dogs and cats is 0.4. The probability they like cats and gerbils is 0.1, and the probability they like gerbils and dogs is 0.2. Finally, the probability they like all three kinds of animal is 0.1.
You can draw all this in a Venn diagram. (Here I've used C for 'likes cats', D for 'likes dogs' and G for 'likes gerbils'.)

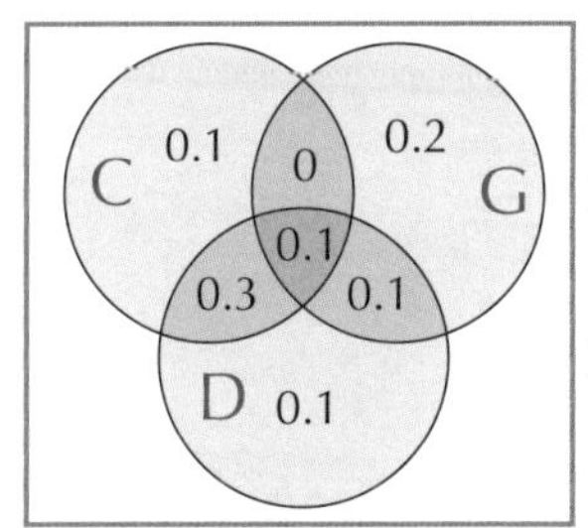

1) Stick in the middle one first — 'likes all 3 animals' (i.e. $C \cap D \cap G$).
2) Then do the 'likes 2 animals' probabilities by taking 0.1 from each of the given 'likes 2 animals' probabilities. (If they like 3 animals, they'll also be in the 'likes 2 animals' bits.)
3) Finally, do the 'likes 1 kind of animal' probabilities, by making sure the total probability in each circle adds up to the probability in the question.

① From the Venn diagram, the probability that someone likes either dogs or cats is 0.7.

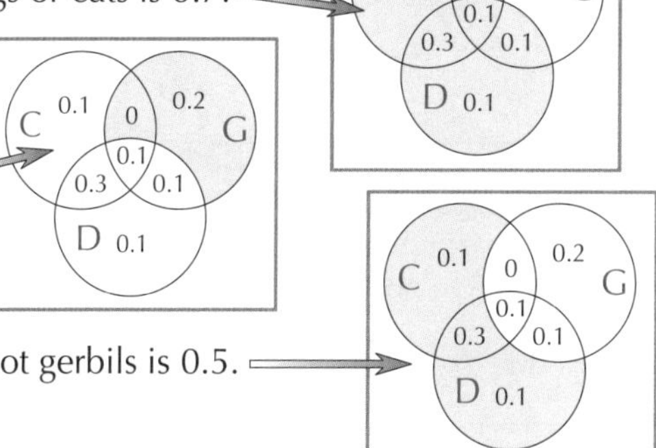

② The probability that someone likes gerbils but not dogs is 0.2.

③ The probability that someone likes cats and dogs, but not gerbils is 0.5.

Practice Questions

1. *A dice and a coin are thrown and the outcomes recorded.*
If a head is thrown, the score on the dice is doubled. If a tail is thrown, 4 is added to the score on the dice.
- ***a) Represent this by means of a sample space diagram.***
- ***b) What is the probability that you score more than 5?***
- ***c) If you throw a tail, what is the probability that you get an even score?***

2. *Half the students in a sixth form college eat sausages for dinner and 20% eat chips.*
10% of those who eat chips also eat sausages. By use of a Venn diagram or otherwise, find:
- ***a) the percentage of students who eat both chips and sausages,***
- ***b) the percentage of students who eat chips but not sausages,***
- ***c) the percentage of students who eat either chips or sausages but not both.***

Sample Exam question:

3. A soap company asked 120 people about the types of soap (from Brands A, B and C) they bought. Brand A was bought by 40 people, Brand B by 30 people and Brand C by 25. Both Brands A and B (and possibly C as well) were bought by 8 people, B and C (and maybe A) were bought by 10 people, and A and C (and maybe B) by 7 people. All three brands were bought by 3 people.

(a) Represent this information in a Venn diagram. [5 marks]

(b) If a person is selected at random, find the probability that:
- (i) they buy at least one of the soaps. [2 marks]
- (ii) they buy least two of the soaps, [2 marks]
- (iii) they buy soap B, given that they buy only one type of soap. [3 marks]

Two heads are better than one — though only half as likely using two coins...

I must admit — I kind of like these pages. This stuff isn't too hard, and it's really useful for answering loads of questions. And one other good thing is that Venn diagrams look, well, nice somehow. But more importantly, when you're filling one in, the thing to remember is that you usually need to 'start from the inside and work out'.

Probability

If you're doing AQA A S1, this stuff is in the Methods module.

So far so good. But I can see you want more.

Mutually Exclusive Events Have No Overlap

If two events can't both happen at the same time (i.e. $P(A \cap B) = 0$) they're called mutually exclusive (or just 'exclusive').

If A and B are exclusive, then the probability of A or B is: $P(A \cup B) = P(A) + P(B)$.

Use the formula from page 18, but put $P(A \cap B) = 0$.

More generally,

For n exclusive events (i.e. only one of them can happen at a time):

$$P(A_1 \cup A_2 \cup ... \cup A_n) = P(A_1) + P(A_2) + ... + P(A_n)$$

Example: Find the probability that a card pulled at random from a pack of cards (no jokers) is either a picture card (a Jack, Queen or King) or the 7, 8 or 9 of clubs.

Call event A — 'I get a picture card', and event B — 'I get the 7, 8 or 9 of clubs'.

Events A and B are mutually exclusive — they can't both happen. Also, $P(A) = \frac{12}{52} = \frac{3}{13}$ and $P(B) = \frac{3}{52}$.

So the probability of either A or B is: $P(A \cup B) = P(A) + P(B) = \frac{12}{52} + \frac{3}{52} = \frac{15}{52}$

The Complement of 'Event A' is 'Not Event A'

An event A will either happen or not happen. The event 'A doesn't happen' is called the complement of A (or A'). On a Venn diagram, it would look like this (because $A \cup A' = S$, the sample space):

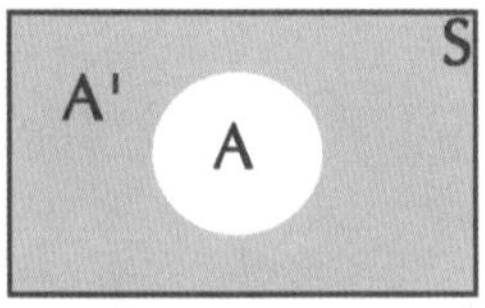

At least one of A and A' has to happen, so...

$$P(A) + P(A') = 1 \quad or \quad P(A') = 1 - P(A)$$

Example: A teacher keeps socks loose in a box. One morning, he picks out a sock. He quickly calculates that the probability of then picking out a matching sock is 0.56. What is the probability of him not picking a matching sock?

Call event A 'picks a matching sock'. Then A' is 'doesn't pick a matching sock'. Now A and A' are complementary events (and P(A) = 0.56), so P(A) + P(A') = 1, and therefore P(A') = 1 – 0.56 = 0.44

Tree Diagrams Show Probabilities for Two or More Events

Each 'chunk' of a tree diagram is a trial, and each branch of that chunk is a possible outcome. Multiplying probabilities along the branches gives you the probability of a series of outcomes.

Example: If Susan plays tennis one day, the probability that she'll play the next day is 0.2. If she doesn't play tennis, the probability that she'll play the next day is 0.6. She plays tennis on Monday. What is the probability she plays tennis:

(i) on both the Tuesday and Wednesday of that week?
(ii) on the Wednesday of the same week?

Let T mean 'plays tennis' (and then T' means 'doesn't play tennis').

Notice that these add up to 1.

(i) Then the probability that she plays on Tuesday and Wednesday is P(T and T) = 0.2 × 0.2 = 0.04 (multiply probabilities since you need a series of outcomes — T and then T).

(ii) Now you're interested in either P(T and T) or P(T' and T). To find the probability of one event or another happening, you have to add probabilities: P(plays on Wednesday) = 0.04 + 0.48 = 0.52.

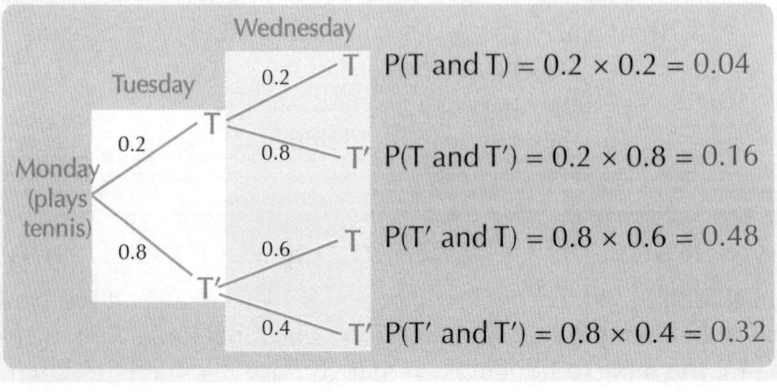

Probability

Sometimes a Branch is Missing

Example:

A box of biscuits contains 5 chocolate biscuits and 1 lemon biscuit.
George takes out three biscuits at random, one at a time, and eats them.

a) Find the probability that he eats 3 chocolate biscuits.

b) Find the probability the last biscuit is chocolate.

Let C mean 'picks a chocolate biscuit' and L mean 'picks the lemon biscuit'

1st pick — 2nd pick — 3rd pick

- C ($\frac{5}{6}$) → C ($\frac{4}{5}$) → C ($\frac{3}{4}$): $\frac{5}{6}\times\frac{4}{5}\times\frac{3}{4}$
- C ($\frac{5}{6}$) → C ($\frac{4}{5}$) → L ($\frac{1}{4}$): $\frac{5}{6}\times\frac{4}{5}\times\frac{1}{4}$
- C ($\frac{5}{6}$) → L ($\frac{1}{5}$) → C (1): $\frac{5}{6}\times\frac{1}{5}\times 1$
- L ($\frac{1}{6}$) → C (1) → C (1): $\frac{1}{6}\times 1\times 1$

After the lemon biscuit there are only chocolate biscuits left, so the tree diagram doesn't 'branch' after an 'L'.

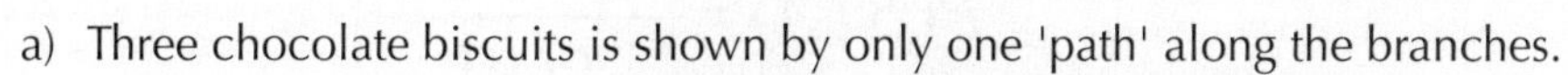

a) Three chocolate biscuits is shown by only one 'path' along the branches.

$$P(\text{C and C and C}) = \frac{5}{6}\times\frac{4}{5}\times\frac{3}{4} = \frac{60}{120} = \frac{1}{2}$$

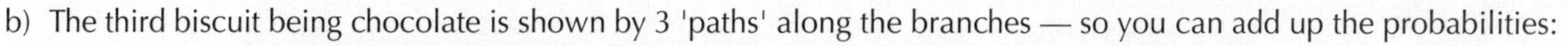

b) The third biscuit being chocolate is shown by 3 'paths' along the branches — so you can add up the probabilities:

$$P(\text{third biscuit is chocolate}) = \left(\frac{5}{6}\times\frac{4}{5}\times\frac{3}{4}\right)+\left(\frac{5}{6}\times\frac{1}{5}\times 1\right)+\left(\frac{1}{6}\times 1\times 1\right) = \frac{1}{2}+\frac{1}{6}+\frac{1}{6} = \frac{5}{6}$$

There's a quicker way to do this, since there's only one outcome where the chocolate isn't picked last:

$$P(\text{third biscuit is not chocolate}) = \frac{5}{6}\times\frac{4}{5}\times\frac{1}{4} = \frac{1}{6}, \text{ so } P(\text{third biscuit is chocolate}) = 1-\frac{1}{6} = \frac{5}{6}$$

Working out the probability of the complement of the event you're interested in is sometimes easier.

Practice Questions

***1.** Arabella rolls two dice and adds the two results together.*

a) What is the probability that she scores a prime number?

b) What is the probability that she scores a square number?

c) What is the probability that she scores a number that is either a prime number or a square number?

***2.** In a school orchestra (made up of pupils in either the upper or lower school), 40% of the musicians are boys. Of the boys, 30% are in the upper school. Of the girls in the orchestra, 50% are in the upper school.*

a) Draw a tree diagram to show the various probabilities.

b) Find the probability that a musician chosen at random is in the upper school.

Sample Exam question:

3. A box contains counters of various colours. There are 3 red counters, 4 white counters and 5 green counters. Two random counters are removed from the jar one at a time. Once removed, the colour of the counter is noted. The first counter is not replaced before the second one is drawn.

(a) Draw a tree diagram to show the probabilities of the various outcomes. [4 marks]

(b) Find the probability that the second disc is green. [2 marks]

(c) Find the probability that both the discs are red. [2 marks]

(d) Find the probability that the two discs are not both the same colour. [3 marks]

Useful quotes: I can live for two months on a good compliment...*

Tree diagrams are another one of those things that are fairly easy to get your head round, but at the same time, are incredibly useful. And if you get stuck trying to work out a probability, it's worth checking to see if the probability of the complementary event would be easier to find — because if you can find one, then you can easily work out the other.

* Mark Twain

Conditional Probability

If you're doing AQA A S1, this stuff is in the Methods module.

Examiners love conditional probability — they can't get enough of it. So learn this well...

P(B|A) means Probability of B, given that A has Already Happened

Conditional probability means the probability of something, given that something else has already happened. For example, P(B|A) means the probability of B, given that A has already happened. More tree diagrams...

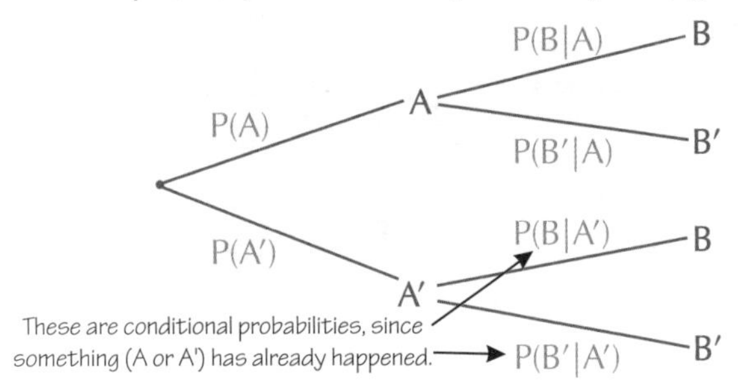

If you multiply probabilities along the branches, you get:

i.e. P(A and B) ⟹ $P(A \cap B) = P(A) \times P(B \mid A)$

You can rewrite this as:

$$P(B \mid A) = \frac{P(A \cap B)}{P(A)}$$

Example: Horace either walks (W) or runs (R) to the bus stop.
If he walks he catches (C) the bus with a probability of 0.3. If he runs he catches it with a probability of 0.7. He walks to the bus stop with a probability of 0.4.
Find the probability that Horace catches the bus.

$$\begin{aligned} P(C) &= P(C \cap W) + P(C \cap R) \\ &= P(W)P(C|W) + P(R)P(C|R) \\ &= (0.4 \times 0.3) + (0.6 \times 0.7) = 0.12 + 0.42 = \underline{0.54} \end{aligned}$$

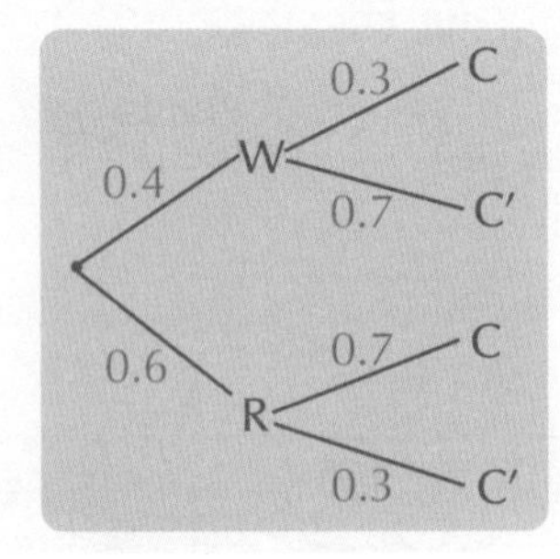

This is easier to follow if you match each part of this working to the probabilities in the tree diagram.

If B is Conditional on A then A is Conditional on B

If B depends on A then A depends on B — and it doesn't matter which event happens first.

Example: Horace turns up at school either late (L) or on time (L′). He is then either shouted at (S) or not (S′).
The probability that he turns up late is 0.4. If he turns up late the probability that he is shouted at is 0.7.
If he turns up on time the probability that he is shouted at is 0.2.
If you hear Horace being shouted at, what is the probability that he turned up late?

1) The probability you want is P(L|S).

Get this the right way round — he's already being shouted at.

2) Use the conditional probability formula: $P(L \mid S) = \dfrac{P(L \cap S)}{P(S)}$

3) The best way to find $P(L \cap S)$ and $P(S)$ is with a tree diagram.

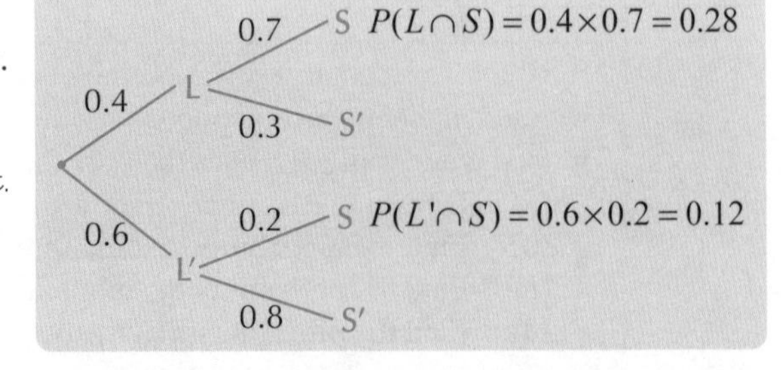

Be careful with questions like this — the information in the question tells you what you need to know to draw the tree diagram with L (or L′) considered first. But you need P(L|S) — where S is considered first. So don't just rush in.

$$P(L \cap S) = 0.4 \times 0.7 = 0.28$$
$$P(S) = P(L \cap S) + P(L' \cap S) = 0.28 + 0.12 = 0.40$$

4) Put these in your conditional probability formula to get:

$$P(L \mid S) = \frac{0.28}{0.4} = 0.7$$

Conditional Probability

Independent Events Have No Effect on Each Other

If the probability of B happening doesn't depend on whether or not A has happened, then A and B are independent.

1) If A and B are independent, $P(A|B) = P(A)$.
2) If you put this in the conditional probability formula, you get: $P(A|B) = P(A) = \frac{P(A \cap B)}{P(B)}$

Or, to put that another way: **For independent events: $P(A \cap B) = P(A)P(B)$**

Example: You are exposed to two infectious diseases — one after the other. The probability you catch the first (A) is 0.25, the probability you catch the second (B) is 0.5, and the probability you catch both of them is 0.2. Are catching the two diseases independent events?

1) You need to compare P(A|B) and P(A) — if they're different, the events aren't independent.

$$P(A|B) = \frac{P(A \cap B)}{P(B)} = \frac{0.2}{0.5} = 0.4 \qquad P(A) = 0.25$$

2) P(A|B) and P(A) are different, so they are not independent.

Practice Questions

1. A and B are two events. $P(A) = 0.4$, $P(B) = 0.3$, $P(A \cap B) = 0.1$.
a) Find $P(B|A)$.
b) Find $P(B|A')$.
c) Say whether or not A and B are independent.

2. Albert eats a limited choice of lunch. He eats either chicken or beef for his main course, and either chocolate pudding or ice cream for dessert. The probability that he eats chicken is 1/3, the probability that he eats ice cream given that he has chicken is 2/5, and the probability that he has ice cream given that he has beef is 3/4.
a) Find the probability he has either chicken or ice cream — but not both.
b) Find the probability that he eats ice cream.
c) Find the probability that he had chicken given that you see him eating ice cream.

Sample Exam questions:

3. V and W are independent events, where $P(V) = 0.2$ and $P(W) = 0.6$.
   (a) Find: (i) $P(V \cap W)$, and (ii) $P(V \cup W)$. [3 marks]
   (b) If U is the event that neither V or W occurs, find $P(U|V')$. [3 marks]

4. For a particular biased dice, the event 'throw a 6' is called event B. $P(B) = 0.2$.
   This biased dice and a fair dice are rolled together. Find the probability that:
   (a) the biased dice doesn't show a 6, [1 mark]
   (b) at least one of the dice shows a 6, [2 marks]
   (c) exactly one of the dice shows a 6, given that at least one of them shows a 6. [3 marks]

Statisticians say: P(Having cake ∩ Eating it) = 0...

It's very easy to assume that events are independent and use the $P(A \cap B)$ formula in the red box on this page — when in fact they're not independent at all and you should be using the formula in the blue box on page 22. I admit that using the wrong formula may make the calculations slightly easier — but think how virtuous you'll feel if you do things the hard way.

Permutations and Combinations

Skip these two pages if you're doing Edexcel S1, AQA A S1 or AQA B S1.

In Permutations, the Order Matters

Permutations and combinations are all about the numbers of ways you can arrange things.

In a permutation, the order matters — so '123' is a different permutation to '321'.

The number of permutations of n distinct objects is n!, where n! = n × (n – 1) × (n – 2) × ... × 2 × 1.

Example: How many different permutations are there of the numbers 5, 7 and 9?

1) The way to do these is to think: 'How many choices do I have for the first number?', 'How many choices do I then have for the second number?' and so on.
2) You can put either 5, 7 or 9 in the first position, so you have 3 choices — I'll choose 7.
3) Then you can put either 5 or 9 in the second position, so you have 2 choices — I'll choose 9.
4) Then I have only 1 choice for what goes in the third position — I have to choose 5.
   So there are 3 × 2 × 1 = 6 permutations of 5, 7 and 9.

Sometimes you only want to arrange a few of the available objects.

The number of permutations of r objects selected from n distinct objects is $\frac{n!}{(n-r)!}$.

This is sometimes written ^nP_r.

Example: How many permutations of 5 letters can you make from the letters of the alphabet (if you don't use any letter twice)?

You have 26 choices for the first letter, 25 for the second, then 24, 23 and 22 for the third, fourth and fifth.

This gives a total of 26 × 25 × 24 × 23 × 22 = 7 893 600 permutations.

This is $\frac{26!}{21!}$.

The number of permutations is Less if some objects are the Same

Say you have *n* objects, but they're not all different — there's a group of *r* that are identical, and another group of *s* that are identical. Then the number of permutations is:

$$\frac{n!}{r! \times s!}$$

Example: How many different arrangements of the letters in the word MISSISSIPPI are there?

There are 11 letters in total — including 4 S's, 4 I's and 2 P's.

So the number of arrangements is: $\frac{11!}{4! \times 4! \times 2!} = 34650$

When the Order Doesn't Matter it's a Combination

In a combination, the order of the objects doesn't matter — so '123' is the same combination as '231'.

The number of combinations of r objects selected from n distinct objects is ^nC_r or $\binom{n}{r}$, where $^nC_r = \binom{n}{r} = \frac{n!}{r!(n-r)!}$

Example: How many ways are there to choose a team of 11 players from a squad of 16?

Easy — just use the formula: $\binom{16}{11} = \frac{16!}{5! \times 11!} = 4368$

Notice that $\binom{16}{11} = \binom{16}{5}$.

Example: How many ways are there to choose 6 lottery numbers from a possible 49?

Use the formula again: $\binom{49}{6} = \frac{49!}{6! \times 43!} = 13\,983\,816$

Cancel the 43! with the rest of the 49! to get $\frac{49!}{6!43!} = \frac{49 \times 48 \times 47 \times 46 \times 45 \times 44}{6 \times 5 \times 4 \times 3 \times 2 \times 1}$

Permutations and Combinations

Use Permutations and Combinations to Find Probabilities

This is a bit like the stuff on page 18 — you count the outcomes that you're interested in, and divide by the total number of possible outcomes. Use the 'permutations and combinations' formulas to do the counting.

Example: I've bought a Lotto ticket — what's the probability that I win the jackpot?

I'm only interested in 1 outcome (the one where my ticket wins). But there are 13983816 possibles (see p24).

So the probability that I win is $\frac{1}{13983816}$.

Example: Six people are to be sat in a row of seats at random. Three of the people are the Bills family. What's the probability that the Bills family are sat together?

1) This is 'number of ways the Bills family are together ÷ the total number of seating arrangements'.
2) The total number of seating arrangements is 6! = 720.
3) Now assume that the Bills are together by pretending they're one person — now you have only 4 people to arrange.
4) These 4 people can be arranged in 4! = 24 ways.
5) However for each of these 24 ways, there are 3! = 6 ways to arrange the Bills family.
   So the number of arrangements with the Bills together is actually 24 × 6 = 144.
6) Therefore the probability that the Bills are together is: $\frac{144}{720} = 0.2$

Practice Questions

1. How many arrangements of the letters in the word statistics are there?

2. Three married couples are about to sit on the 6 back seats of a coach.
- ***a) How many different arrangements of the 6 people are possible?***
- ***b) In how many of these is Mr Brown sitting next to his wife?***
- ***c) What is the probability that, from a randomly chosen arrangement, Mr Brown is not sitting next to his wife?***

3. The letters of the alphabet are written on 26 identical discs and placed in a bag. Six discs are then taken at random.
- ***a) How many different sets of 6 are possible?***
- ***b) What is the probability that 5 vowels are taken?***
- ***c) What is the probability that at least 1 vowel is taken?***
- ***d) If you select the letters A, B, R, I, O and Z, and arrange 4 of them randomly, what is the probability that your arrangement starts with a vowel?***

Sample Exam question:

4. A group of 5 men and 5 women stand in a line to have their photos taken.
   (a) In how many different ways can they stand in line? [1 mark]
   (b) What is the probability that no two men and no two women stand next to each other? [3 marks]
   (c) Six people from the group are to pose for an additional photograph. At least 2 of the people chosen must be men. In how many different ways can the 6 people be chosen? [3 marks]

Maths proverb: Absence of stats makes the heart grow fonder...

Now you have to promise not to say 'combination' when you really mean 'permutation', okay. The "*How many choices?*" approach to finding the number of permutations is a good one to remember. But it's definitely worth committing the ^nP_r and ^nC_r formulas to memory as well — they can save you having to actually think. At the first sight of a "*Choose r from n*" question, you just need to decide whether the order of the things matters, and then bung the numbers in the right formula.

Probability Distributions

DON'T SKIP these pages — you need to know this stuff no matter what syllabus you're doing.

This stuff isn't hard — but it can seem a bit weird at times.

*Getting your head round this **Boring Stuff** will help a bit*

This first bit isn't particularly interesting. But understanding the difference between X and x (bear with me) might make the later stuff a bit less confusing. Might.

1) X (upper case) is just the name of a random variable. So X could be 'score on a dice' — it's just a name.
2) A random variable doesn't have a fixed value. Like with a dice score — the value on any 'roll' is all down to chance.
3) x (lower case) is a particular value that X can take. So for one roll of a dice, x could be 1, 2, 3, 4, 5 or 6.

4) Discrete random variables only have a certain number of possible values. Often these values are whole numbers, but they don't have to be. Usually there are only a few possible values (e.g. the possible scores with one roll of a dice).
5) The probability density function (pdf) is a list of the possible values of x, plus the probability for each one.

*All the Probabilities **Add up to 1***

For a discrete random variable X:

$$\sum_{\text{all } x} P(X = x) = 1$$

Example: The random variable X has pdf $P(X = x) = kx$ for $x = 1, 2, 3$. Find the value of k.

So X has three possible values ($x = 1$, 2 and 3), and the probability of each is kx (where you need to find k).

It's easier to understand with a table:

x	1	2	3
$P(X = x)$	$k \times 1 = k$	$k \times 2 = 2k$	$k \times 3 = 3k$

Now just use the formula: $\sum_{\text{all } x} P(X = x) = 1$ Here, this means: $k + 2k + 3k = 6k = 1$

i.e. $k = \frac{1}{6}$

Piece of cake.

The **mode** is the most likely value — so it's the value with the biggest probability.

Example: The discrete random variable X has pdf as shown in the table below.

x	0	1	2	3	4
$P(X = x)$	0.1	0.2	0.3	0.2	a

Find: (i) the value of a, (ii) $P(2 \leq X < 4)$, (iii) the mode

(i) Use the formula $\sum_{\text{all } x} P(X = x) = 1$ again.

From the table: $0.1 + 0.2 + 0.3 + 0.2 + a = 1$
$0.8 + a = 1$
$a = 0.2$

Careful with the inequality signs — you need to include $x = 2$ but not $x = 4$.

(ii) This is asking for the probability that 'X is greater than or equal to 2, but less than 4'. Easy — just add up the probabilities.

$P(2 \leq X < 4) = P(X = 2) + P(X = 3) = 0.3 + 0.2 = 0.5$

(iii) The mode is the value of x with the biggest probability — so mode = 2.

Probability Distributions

Do Complicated questions Bit by bit

Example: A game involves rolling two fair dice. If the sum of the scores is greater than 10 then the player wins 50p. If the sum is between 8 and 10 (inclusive) then he wins 20p. Otherwise he gets nothing. If X is the random variable "amount player wins", find the pdf of X.

There are 3 possible values for X (0, 20 and 50) and you need the probability of each.
To work these out, you need the probability of getting various totals on the dice.

① You need to know $P(8 \le \text{score} \le 10)$ — the probability that the score is between 8 and 10 inclusive (i.e. including 8 and 10) and $P(11 \le \text{score} \le 12)$ — the probability that the score is greater than 10.
This means working out: P(score = 8), P(score = 9), P(score = 10), P(score = 11) and P(score = 12). Use a table...

② Score on dice 1 (columns), Score on dice 2 (rows):

+	1	2	3	4	5	6
1	2	3	4	5	6	7
2	3	4	5	6	7	8
3	4	5	6	7	8	9
4	5	6	7	8	9	10
5	6	7	8	9	10	11
6	7	8	9	10	11	12

There are 36 possible outcomes...
...5 of these have a total of 8 — so the probability of scoring 8 is $\frac{5}{36}$,
...4 have a total of 9 — so the probability of scoring 9 is $\frac{4}{36}$,
...the probability of scoring 10 is $\frac{3}{36}$
...the probability of scoring 11 is $\frac{2}{36}$
...the probability of scoring 12 is $\frac{1}{36}$

③ To find the probabilities you need, you just add the right bits together:

$P(X = 20p) = P(8 \le \text{score} \le 10) = \frac{5}{36} + \frac{4}{36} + \frac{3}{36} = \frac{12}{36} = \frac{1}{3}$ $P(X = 50p) = P(11 \le \text{score} \le 12) = \frac{2}{36} + \frac{1}{36} = \frac{3}{36} = \frac{1}{12}$

To find P(X = 0) just take the total of the two probabilities above from 1 (since X = 0 is the only other possibility).

$P(X = 0) = 1 - \left(\frac{12}{36} + \frac{3}{36}\right) = 1 - \frac{15}{36} = \frac{21}{36} = \frac{7}{12}$

④ Now just stick all this info in a table (and check that the probabilities all add up to 1):

x	0	20	50
$P(X = x)$	$\frac{7}{12}$	$\frac{1}{3}$	$\frac{1}{12}$

Practice Questions

1) The probability density function of Y is shown in the table.
(a) Find the value of k. (b) Find P(Y<2).

y	0	1	2	3
P(Y = y)	0.5	k	k	3k

2) An unbiased six-sided dice has faces marked 1, 1, 1, 2, 2, 3.
The dice is rolled twice. Let X be the random variable "sum of the two scores on the dice".
Show that $P(X = 4) = \frac{5}{18}$. Find the probability density function of X.

Sample exam question:

3) In a game a player tosses three fair coins. If three heads occur then the player gets 20p; if two heads occur then the player gets 10p; otherwise the player gets nothing.
(a) If X is the random variable 'amount received' tabulate the probability density function of X. [4 marks]
The player pays 10p to play one game.
(b) Use the probability density function to find the probability that the player wins (i.e. gets more money than he pays to play) in one game. [2 marks]

Useful quotes: All you need in life is ignorance and confidence, then success is sure...*
Remember I said on page 19 that the 'counting the outcomes' approach was useful — well there you go. And if you remember how to do that, then you can work out a pdf. And if you can work out a pdf, then you can often begin to unravel even fairly daunting-looking questions. But most of all, REMEMBER THAT ALL THE PROBABILITIES ADD UP TO 1. (Ahem.)

* Mark Twain

The Distribution Function

Skip this page if you're doing OCR B S1.

The pdf gives the probability that X will equal this or equal that. The distribution function tells you something else.

'Distribution Function' is the same as 'Cumulative Distribution Function'

The (cumulative) distribution function F(x) gives the probability that X will be less than or equal to a particular value.

$$F(X) = P(X \leq x)$$

Example: The probability density function of the discrete random variable H is shown in the table. Find the cumulative distribution function F(H).

h	0.1	0.2	0.3	0.4
$P(H = h)$	$\frac{1}{4}$	$\frac{1}{4}$	$\frac{1}{3}$	$\frac{1}{6}$

There are 4 values of h, so you have to find the probability that H is less than or equal to each of them in turn. It sounds trickier than it actually is — you only have to add up a few probabilities...

$F(0.1) = P(H \leq 0.1)$ — this is the same as $P(H = 0.1)$, since H can't be less than 0.1. So $F(0.1) = \frac{1}{4}$.

$F(0.2) = P(H \leq 0.2)$ — this is the probability that H = 0.1 or H = 0.2. So $F(0.2) = P(H = 0.1) + P(H = 0.2) = \frac{1}{4} + \frac{1}{4} = \frac{1}{2}$.

$F(0.3) = P(H \leq 0.3) = P(H = 0.1) + P(H = 0.2) + P(H = 0.3) = \frac{1}{4} + \frac{1}{4} + \frac{1}{3} = \frac{5}{6}$.

$F(0.4) = P(H \leq 0.4) = P(H = 0.1) + P(H = 0.2) + P(H = 0.3) + P(H = 0.4) = \frac{1}{4} + \frac{1}{4} + \frac{1}{3} + \frac{1}{6} = 1$.

P(X ≤ largest value of x) is always 1.

Finally, put these values in a table, and you're done...

h	0.1	0.2	0.3	0.4
$F(H) = P(H \leq h)$	$\frac{1}{4}$	$\frac{1}{2}$	$\frac{5}{6}$	1

Sometimes they ask you to work backwards...

Example: The formula below gives the cumulative distribution function F(X) for a discrete random variable X. Find k, and the probability density function.

$F(X) = kx$, for $x = 1, 2, 3$, and 4.

1. First find k. You know that X has to be 4 or less — so $P(X \leq 4) = 1$.

   Put x = 4 into the cumulative distribution function: $F(4) = P(X \leq 4) = 4k = 1$, so $k = \frac{1}{4}$.

2. Now you can work out the probabilities of X being less than 1, 2, 3 and 4.

   $F(1) = P(X \leq 1) = 1 \times k = \frac{1}{4}$, $F(2) = P(X \leq 2) = 2 \times k = \frac{1}{2}$, $F(3) = P(X \leq 3) = 3 \times k = \frac{3}{4}$, $F(4) = P(X \leq 4) = 1$

3. This is the clever bit...

   $P(X = 4) = P(X \leq 4) - P(X \leq 3) = 1 - \frac{3}{4} = \frac{1}{4}$

   Think about it... ...if it's less than or equal to 4, ...but it's not less than or equal to 3, ...then it has to be 4.

   $P(X = 3) = P(X \leq 3) - P(X \leq 2) = \frac{3}{4} - \frac{1}{2} = \frac{1}{4}$

   $P(X = 2) = P(X \leq 2) - P(X \leq 1) = \frac{1}{2} - \frac{1}{4} = \frac{1}{4}$

   $P(X = 1) = P(X \leq 1) = \frac{1}{4}$ — Because x doesn't take any values less than 1.

4. Finish it all off by making a table. The pdf of X is:

x	1	2	3	4
$P(X = x)$	$\frac{1}{4}$	$\frac{1}{4}$	$\frac{1}{4}$	$\frac{1}{4}$

Or you could write it as a formula: $P(X = x) = \frac{1}{4}$ for $x = 1, 2, 3, 4$

Discrete Uniform Distributions

You only need to know about 'discrete uniform distributions' if you're doing Edexcel S1.

When every value of X is equally likely, you've got a uniform distribution. For example, rolling an unbiased dice gives you a discrete uniform distribution. (It's 'discrete' because there are only a few possible outcomes.)

In a Discrete Uniform Distribution the Probabilities are Equal

The pdf of a discrete uniform distribution looks like this — in this version there are only 4 possible values:

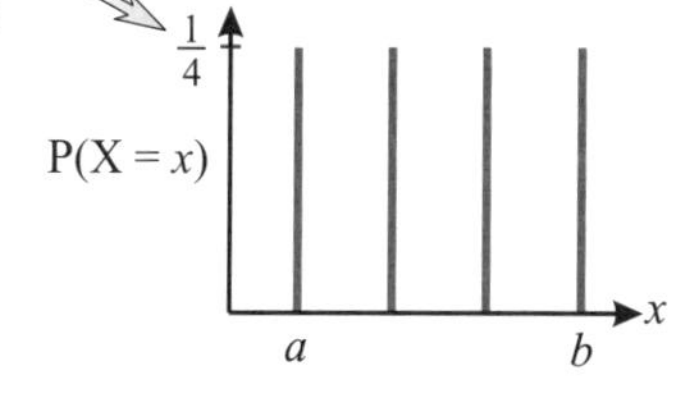

For a discrete uniform distribution X which can take consecutive whole number values a, a+1, a+2,...,b, the mean (or expected value) and variance are easy to work out.

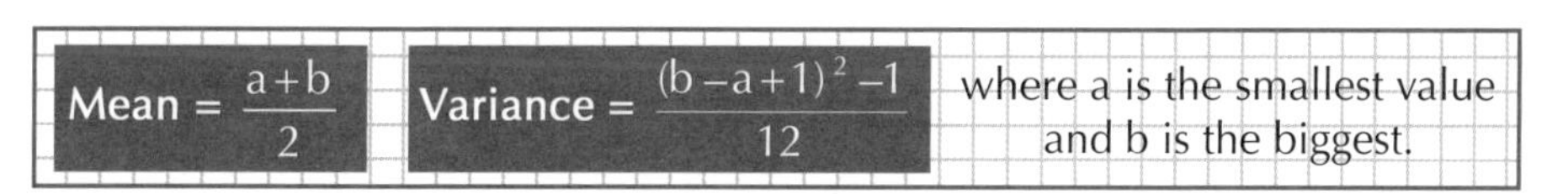

$$\text{Mean} = \frac{a+b}{2} \qquad \text{Variance} = \frac{(b-a+1)^2-1}{12}$$

where a is the smallest value and b is the biggest.

See page 30 for more info about the expected value and variance of a random variable.

Example: Find the mean and variance of the score on an unbiased six-sided dice.

If X is the random variable 'score on a dice', then X has a discrete uniform distribution — like in this table:

x	1	2	3	4	5	6
$P(X = x)$	$\frac{1}{6}$	$\frac{1}{6}$	$\frac{1}{6}$	$\frac{1}{6}$	$\frac{1}{6}$	$\frac{1}{6}$

The symmetry of the distribution should tell you where the mean is — it has to be halfway between 1 and 6.

The smallest value of x is 1 and the biggest is 6 — so a = 1 and b = 6.
Now just stick the numbers in the formulas:

$$\text{Mean} = \frac{a+b}{2} = \frac{1+6}{2} = \frac{7}{2} = \underline{3.5}$$

$$\text{Variance} = \frac{(b-a+1)^2-1}{12} = \frac{(6-1+1)^2-1}{12} = \frac{35}{12} = \underline{2.92} \text{ to 3 sig. fig.}$$

Practice Questions

1) ***The probability density function for the random variable W is given in the table. Find the cumulative distribution function.***

w	**0.2**	**0.3**	**0.4**	**0.5**
P(W = w)	**0.2**	**0.2**	**0.3**	**0.3**

2) ***The cumulative distribution function for a random variable R is given in the table. Calculate the probability density function for R. Find P(0 ≤ R ≤ 1).***

r	**0**	**1**	**2**
F(r) = P(R ≤ r)	**0.1**	**0.5**	**1**

3) ***The discrete random variable X has a uniform distribution, P(X = x) = k for x = 0, 1, 2, 3 and 4. Find the value of k, and then find the mean and variance of X.***

Sample exam questions:

4) A discrete random variable X can only take values 0, 1, 2 and 3. Its pdf is shown in the table.

x	0	1	2	3
$P(X = x)$	$2k$	$3k$	k	k

(a) Find the value of k. [1 mark]
(b) Calculate the distribution function for X. [4 marks]
(c) Calculate P(X > 2). [1 mark]

5) The random variable X takes the values 0, 1, 2, 3, 4, 5, 6, 7, 8 and 9 with equal probability.
(a) Write down the pdf of X. [1 mark]
(b) Find the mean and variance of X. [3 marks]
(c) Calculate the probability that X is less than the mean. [2 marks]

Discreet distributions are more 'British' than those lurid, gaudy ones...

If you've got a pdf, then you can easily work out the distribution function. And if you've got a distribution function, you can work out the pdf, as long as you remember the clever wee trick on page 28. The mean of a discrete uniform distribution is weird. I mean (no stats-pun intended), if you've got a dice, then the expected value is 3½. But I went to a casino recently and bet on 3½ every time at the 'Guess the Dice' table. But not once did the dice ever land on 3½. Not once. I lost loads.

Expected Values, Mean and Variance

Skip this page if you're doing OCR B S1 or AQA B S1.

This is all about the mean and variance of random variables — not a load of data. It's a tricky concept, but bear with it.

The *Mean* of a random variable is the same as the *Expected Value*

You can work out the expected value (or the mean) E(X) of a random variable X.

The expected value (a kind of 'theoretical mean') is what you'd expect the mean of X to be if you took loads of readings. In practice, the mean of your results is unlikely to match the theoretical mean exactly, but it should be pretty near.

If the possible values of X are $x_1, x_2, x_3, \ldots$ then the expected value of X is:

$$\text{Mean} = \text{Expected Value } E(X) = \sum x_i P(X = x_i) = \sum x_i p_i$$

$p_i = P(X = x_i)$

Example: The probability distribution of X, the number of daughters in a family of 3 children, is shown in the table. Find the expected number of daughters.

x_i	0	1	2	3
p_i	$\frac{1}{8}$	$\frac{3}{8}$	$\frac{3}{8}$	$\frac{1}{8}$

$$\text{Mean} = \sum x_i p_i = \left(0\times\tfrac{1}{8}\right)+\left(1\times\tfrac{3}{8}\right)+\left(2\times\tfrac{3}{8}\right)+\left(3\times\tfrac{1}{8}\right) = 0+\tfrac{3}{8}+\tfrac{6}{8}+\tfrac{3}{8} = \tfrac{12}{8} = 1.5$$

So the expected number of daughters is 1.5 — which sounds a bit weird.
But all it means is that if you check a large number of 3-child families, the mean will be close to 1.5.

The *Variance* measures how *Spread Out* the distribution is

You can also find the variance of a random variable. It's the 'expected variance' of a large number of readings.

$$\text{Var}(X) = E(X^2) - [E(X)]^2 = \sum x_i^2 p_i - \left(\sum x_i p_i\right)^2$$

This formula needs $E(X^2) = \sum x_i^2 p_i$ — take each possible value of x, square it, multiply it by its probability and then add up all the results.

Example: Work out the variance for the '3 daughters' example above:

First work out $E(X^2)$: $E(X^2) = \sum x_i^2 p_i = \left(0^2\times\tfrac{1}{8}\right)+\left(1^2\times\tfrac{3}{8}\right)+\left(2^2\times\tfrac{3}{8}\right)+\left(3^2\times\tfrac{1}{8}\right)$
$= 0+\tfrac{3}{8}+\tfrac{12}{8}+\tfrac{9}{8} = \tfrac{24}{8} = \underline{3}$

Now you take away the mean squared: $\text{Var}(X) = E(X^2) - [E(X)]^2 = 3 - 1.5^2 = 3 - 2.25 = \underline{0.75}$

Example: X has the probability density function $P(X = x) = k(x + 1)$ for $x = 0, 1, 2, 3, 4$.
Find the mean and variance of X.

1. First you need to find k — work out all the probabilities and make sure they add up to 1.
   $P(X = 0) = k \times (0 + 1) = k$. Similarly, $P(X = 1) = 2k$, $P(X = 2) = 3k$, $P(X = 3) = 4k$, $P(X = 4) = 5k$.

   So $k + 2k + 3k + 4k + 5k = 1$, i.e. $15k = 1$, and so $k = \frac{1}{15}$

   Now you can work out $p_1, p_2, p_3, \ldots$ where $p_1 = P(X = 1)$ etc.

2. Now use the formulas — find the mean E(X) first:

$$E(X) = \sum x_i p_i = \left(0\times\tfrac{1}{15}\right)+\left(1\times\tfrac{2}{15}\right)+\left(2\times\tfrac{3}{15}\right)+\left(3\times\tfrac{4}{15}\right)+\left(4\times\tfrac{5}{15}\right) = \tfrac{40}{15} = \underline{\tfrac{8}{3}}$$

For the variance you need $E(X^2)$:

$$E(X^2) = \sum x_i^2 p_i = \left(0^2\times\tfrac{1}{15}\right)+\left(1^2\times\tfrac{2}{15}\right)+\left(2^2\times\tfrac{3}{15}\right)+\left(3^2\times\tfrac{4}{15}\right)+\left(4^2\times\tfrac{5}{15}\right) = \tfrac{130}{15} = \underline{\tfrac{26}{3}}$$

And finally: $Var(X) = E(X^2) - [E(X)]^2 = \frac{26}{3} - \left(\frac{8}{3}\right)^2 = \underline{\frac{14}{9}}$

Expected Values, Mean and Variance

Skip 'functions of random variables' if you're doing OCR A S1 or OCR B S1.

You can use the *Expected Value* and *Variance* formulas for *Functions*

$$E(aX + b) = aE(X) + b \qquad Var(aX + b) = a^2Var(X)$$

Here a and b are any numbers.

Example: If E(X) = 3 and Var(X) = 7, find E(2X+5) and Var(2X+5).

Easy. $E(2X + 5) = 2E(X) + 5 = (2 \times 3) + 5 = 11$

$Var(2X + 5) = 2^2Var(X) = 4 \times 7 = 28$

Example: The discrete random variable X has the following probability distribution:

x	2	3	4	5	6
$P(X = x)$	0.1	0.2	0.3	0.2	k

Find: a) k, b) E(X), c) Var(X), d) E(3X – 1), e) Var(3X – 1)

Slowly, slowly — one bit at a time...

a) Remember the probabilities add up to 1 — $0.1 + 0.2 + 0.3 + 0.2 + k = 1$, and so $k = 0.2$

b) Now you can use the formula to find E(X): $E(X) = \sum x_i p_i = (2\times0.1)+(3\times0.2)+(4\times0.3)+(5\times0.2)+(6\times0.2) = 4.2$

c) Next work out E(X²): $E(X^2) = \sum x_i^2 p_i = (2^2\times0.1)+(3^2\times0.2)+(4^2\times0.3)+(5^2\times0.2)+(6^2\times0.2) = 19.2$

and then the variance is easy: $Var(X) = E(X^2) - [E(X)]^2 = 19.2 - 4.2^2 = 1.56$

d) You'd expect the question to get harder but it doesn't: $E(3X - 1) = 3E(X) - 1 = 3 \times 4.2 - 1 = 11.6$

e) And finally: $Var(3X - 1) = 3^2Var(X) = 9 \times 1.56 = 14.04$

Practice Questions

1) A discrete random variable X has the probability distribution shown in the table, where k is a constant.

x_i	1	2	3	4
p_i	$\frac{1}{6}$	$\frac{1}{2}$	k	$\frac{5}{24}$

a) Find the value of k.
b) Find E(X) and show Var(X) = 63/64
c) Find E(2X – 1) and Var(2X – 1)

Sample exam question:

2) A discrete random variable X has the pdf $P(X = x) = ax$ for $x = 1, 2, 3$, where a is a constant.

(a) Show $a = \frac{1}{6}$. [1 mark]

(b) Find E(X). [2 marks]

(c) If $Var(X) = \frac{5}{9}$ find $E(X^2)$. [2 marks]

(d) Find E(3X + 4) and Var(3X + 4). [3 marks]

Statisticians say: E(Bird in hand) = E(2 Birds in bush)...

The mean and variance here are theoretical values — don't get them confused with the mean and variance of a load of practical observations. This 'theoretical' variance has a similar formula to the variance formula on p8, though — it's just "E(X-squared) minus E(X)-squared". And you can still take the square root of the variance to get the standard deviation.

The Binomial Distribution

Skip this page if you're doing Edexcel S1.

With the binomial distribution (as with so much in Stats), the notation and jargon can make it seem really hard... ...but once you've practised a few questions, you realise it's about as hard as a really squishy sponge.

The *Binomial Distribution* is all about *Success and Failure*

Binomial distributions are for finding out the probability of a certain number of 'successes' in a fixed number of 'trials'.

To use the Binomial distribution, look for the following:

1) **A fixed number of trials (n), e.g. 12 eggs in a box, 10 rolls of a dice...**
2) **Only two outcomes of each trial.** These are called "success" or "failure", but don't necessarily sound like success and failure, e.g. success = the egg is cracked, failure = the egg is not cracked; success = a six is rolled, failure = a six is not rolled.
3) **Each trial is independent of all others, e.g. the first egg being broken does not affect the other eggs being broken. The probability of success (p) is constant for each trial.**

If the random variable X has a Binomial distribution, then we write $X \sim B(n, p)$

Sometimes you need to use the *Probability Formula*

The probability that X takes a particular value x is:

$$P(X = x) = \binom{n}{x} p^x q^{n-x}, \text{ where } q = 1 - p$$

and $\binom{n}{x} = \frac{n!}{x!(n-x)!}$ is a binomial coefficient.

Use a calculator's ^nC_r button to work these out, e.g. 5 ^nC_r 3 finds $\binom{5}{3}$.

Example: Eggs are packed in boxes of 12. The probability that each egg is broken is 0.35. Find the probability that in a random box of eggs:
(i) there are 4 broken eggs, (ii) there are less than 3 broken eggs.

Let X be the random variable "number of broken eggs". Here n = 12 (there are 12 eggs in the box, so 12 trials). Also $p = 0.35$ (and so $q = 1 - 0.35 = 0.65$) which means that $X \sim B(12, 0.35)$

(i) First you need the probability of 4 'successes', i.e. X = 4. Just use the formula:

Here, 'success' means the egg is broken and 'failure' means the egg is intact.

$$P(X=4) = \binom{12}{4} \times 0.35^4 \times 0.65^8 = 495 \times 0.35^4 \times 0.65^8 = 0.237 \text{ to 3 significant figures}$$

(ii) Then you need the probability that X is less than 3 — so X could be 0, 1 or 2. Add up the separate probabilities.

$$P(X<3) = P(X=0) + P(X=1) + P(X=2)$$
$$= \binom{12}{0} \times 0.35^0 \times 0.65^{12} + \binom{12}{1} \times 0.35^1 \times 0.65^{11} + \binom{12}{2} \times 0.35^2 \times 0.65^{10}$$
$$= (1 \times 1 \times 0.005688) + (12 \times 0.35 \times 0.008751) + (66 \times 0.1225 \times 0.01346) = 0.151 \text{ to 3 sig. fig.}$$

For a Binomial Distribution, *Mean = np* and *Variance = npq*

Yep, I'll say that again...

If $X \sim B(n, p)$, Mean or Expected Value $E(X) = np$
Variance, $Var(X) = np(1 - p) = npq$

Example: What is the mean and variance of X, if X is the number of sixes when I throw a fair dice 100 times?

There are a fixed number of independent trials (= 100) and the probability of success each time is the same (= $\frac{1}{6}$).

So this is a binomial distribution — $X \sim B\left(100, \frac{1}{6}\right)$

Mean = Expected value $E(X) = np = 100 \times \frac{1}{6} = 16.7$

Variance = $Var(X) = np(1-p) = 100 \times \frac{1}{6} \times \frac{5}{6} = 13.9$

The Binomial Distribution

Skip the stuff about 'binomial tables' if you're doing Edexcel S1.

It's Quicker to use Tables when you can

Example: A fair dice is rolled 10 times. Find the probability that: a) there are less than 5 sixes rolled,
b) there are 3 sixes rolled,
c) there are more than 5 sixes rolled.

Let X be the random variable "Number of sixes rolled". Then $X \sim B\left(10, \frac{1}{6}\right)$.

a) Look in your book of tables for the cumulative binomial tables.
These give probabilities for $X \leq x$ (less than or equal to). You need $P(X < 5) = P(X = 0,1,2,3 \text{ or } 4) = P(X \leq 4)$.
Find the table for n = 10. Look down the left-hand side until you reach x = 4.
Go across the row of figures until you're in the column with p = 1/6. Answer: P(X < 5) = 0.9845.

b) $P(X = 3) = P(X \leq 3) - P(X \leq 2)$
Look these up in the tables.
[P (X = 0, 1, 2 or 3) – P (X = 0, 1 or 2)]. This gives P(X = 3) = 0.9303 – 0.7752 = 0.1551

c) $P(X > 5) = 1 - P(X \leq 5)$.
i.e. P (X = 6, 7, 8, 9 or 10) = 1 – P (X = 0, 1, 2, 3, 4 or 5) = 1 – 0.9976 = 0.0024

Find the Number of 'Failures' if you have to

Binomial tables usually just go up to p = 0.5. If p > 0.5, you have to look up q in the tables instead.

Example: If X ~ B(8, 0.6), find P(X = 3).

The question's asking for the probability of 3 successes in 8 trials, where P(success) = 0.6.
But the tables don't help, since p is too big.
Instead, you'll have to find the probability of 5 failures, where the probability of failure q = 0.4.

If Y is the number of failures, then Y ~ B(8, 0.4). You need P(Y = 5).
$P(Y = 5) = P(Y \leq 5) - P(Y \leq 4) = 0.9502 - 0.8263 = 0.1239$

Practice Questions

1) The random variable X has a binomial distribution with n = 5 and p = 0.3.
Calculate (a) P(X = 2) (b) P(X ≤ 3) (c) P(X < 2) (d) E(X) (e) Var(X)

2) A fair dice is rolled 20 times. Find the probability that the number of sixes is (a) exactly 10, (b) at least 10.

3) 30% of the meals served in a canteen contain a salad. If 10 customers in the canteen are selected at random, what is the probability that fewer than 3 of them have a salad?

Sample exam question:

4) In a room of 30 people, the probability that each person was born on a Saturday is 1/7.
(a) State the distribution of the number of people born on a Saturday and the conditions needed for this distribution. [5 marks]
(b) Calculate the probability that exactly 5 of the people in the room were born on a Saturday. [2 marks]

Binomial distributions are much like exams — all about success and failure...

Binomial distributions come up loads, so learning to love them is probably the best advice I could give you. There are 3 formulas you need to use — the one for a probability, plus those for the mean and variance. Then you need to get used to using the cumulative binomial tables — they tell you the probability that X is less than any value. The tables are slightly different depending on which Exam board you're doing, so have a practice with the actual tables you'll be using come the real Exam. Oh, and one more thing... Exam boards like you to quote your answers from tables in full — so don't round.

The Geometric Distribution

You only need these two pages if you're doing OCR A S1.

Imagine playing a board game where you need to throw a six to start. If you're lucky, you'll get a six first throw. If you're unlucky, you might have to wait a while. But if you're having one of those days, it could take forever...

The Geometric Distribution is about Waiting for Success

Now then... suppose X is the number of throws it takes you to get a six, then X can take the values 1, 2, 3, 4, 5, 6, 7, 8...

The probability of getting a six with the first throw, $P(X = 1) = \frac{1}{6}$

The probability of getting a six with the second throw, $P(X = 2) = \frac{5}{6} \times \frac{1}{6}$

Because you get 'not a six' first, and then you get a six.

The probability of getting a six with the third throw, $P(X = 3) = \frac{5}{6} \times \frac{5}{6} \times \frac{1}{6} = \left(\frac{5}{6}\right)^2 \times \frac{1}{6}$

More generally, the probability of needing r throws to get a six, $P(X = r) = \left(\frac{5}{6}\right)^{r-1} \times \frac{1}{6}$

This is what geometric distributions are all about — how many attempts it takes to achieve 'success'.

Here $X \sim \text{Geo}(\frac{1}{6})$ — meaning that X follows a geometric distribution, and the probability of success each time is $\frac{1}{6}$.

Learn the Geometric Distribution Formula

If $X \sim \text{Geo}(p)$ then the probability that it takes r goes to gain a success is:

$$P(X = r) = (1 - p)^{r-1} \times p$$

Just like with the binomial distribution, the events need to be independent and have a fixed probability of success (p) and failure ($1 - p$). A big difference though is that the number of trials can go on for ever and ever and ever...

Example: $X \sim \text{Geo}(0.2)$.
Find: a) $P(X = 3)$, b) $P(X \leq 3)$, c) $P(X > 3)$

a) Use the formula — that's all you need to do...

$P(X = 3) = (1 - 0.2)^{3-1} \times (0.2) = 0.8^2 \times 0.2 = 0.128$

b) This bit's not too bad either — you need to add some probabilities together:

$P(X \leq 3) = P(X = 1) + P(X = 2) + P(X = 3) = 0.2 + (0.8 \times 0.2) + (0.8^2 \times 0.2) = 0.488$

c) If X is greater than 3, then it's not less than or equal to 3 — so subtract the previous probability from 1:

$P(X > 3) = 1 - 0.488 = 0.512$

Example: During May the probability that it will rain on a given day is 0.15.
Find the probability that the first rainy day in May is May 5th.
(Assume the probability of rain each day is independent of what happens on other days.)

You're waiting for something to happen — must be a geometric distribution.
'Success' here means a rainy day, so $X \sim \text{Geo}(0.15)$.
And you need the probability that X is 5. Use the formula:

$P(X = 5) = (1 - 0.15)^{5-1} \times (0.15) = 0.85^4 \times 0.15 = 0.0783$ to 3 sig. fig.

The Geometric Distribution

Déjà vu. If you can do it with the binomial distribution then this geometric stuff will be a breeze.

The Mean is the Expected Number of Goes it takes to get a Success

$$\text{If } X \sim \text{Geo}(p) \text{ then } E(X) = \frac{1}{p}$$

So if you're playing Ludo and you need a six to start, on average it'll take $\frac{1}{\left(\frac{1}{6}\right)} = 6$ tries to get it.

Example: If you play once a week, how long would you expect it to take to win the Lotto jackpot?

First you need the probability of success:

$$P(\text{win the jackpot}) = \frac{6\times5\times4\times3\times2\times1}{49\times48\times47\times46\times45\times44} = \frac{1}{13\,983\,816}$$ — so $X \sim \text{Geo}\left(\frac{1}{13\,983\,816}\right)$

So you'd expect to have to wait 13 983 816 weeks — or about 269 000 years.

Example: A student is waiting for a bus. Past experience shows that 2% of vehicles in the town are buses.

a) Write down the mean number of vehicles up to and including the first bus.
b) Calculate (to 3 sig. fig.) the probability that the first bus is the 8th vehicle to arrive.
c) Calculate the probability that there is at least one bus among the first 8 vehicles.

The student's waiting for something to happen, so it's a geometric distribution. In fact, $X \sim \text{Geo}(0.02)$.

a) The mean is $\frac{1}{0.02} = 50$

b) Now you need the probability that the first success is on the 8th trial — use the standard formula:

$$P(X = 8) = (1 - 0.02)^{8-1} \times 0.02 = 0.0174 \text{ (to 3 sig. fig.)}$$

c) Careful — you've got a fixed number of events now, so this needs a binomial distribution.
Let the random variable Y be 'the number of buses in the first 8 vehicles'. Then $Y \sim B(8, 0.02)$.
P(there is at least one bus in the first 8 vehicles) = 1 – P(no buses in the first 8 vehicles).
This is $1 - (0.98)^8 = 1 - 0.851 = 0.149$ (to 3 sig. fig.)

Practice Questions

1) $X \sim \text{Geo}(0.1)$. Find the probability that it takes 15 attempts to record a 'success'. What is the expected value of X?

Sample exam question:

2) A game involves throwing 2 dice and gaining a double to start.
(a) Find the mean number of throws needed to start the game. [2 marks]
(b) Calculate (to 3 significant figures) the probability that it takes:
(i) 4 throws, [2 marks]
(ii) at least 5 throws. [3 marks]
(c) Two players are about to play the game. Find the probability that after they have each had 4 throws of the dice at least one of them has not started the game. [2 marks]

So I should buy an extra Lottery ticket then, you reckon...

Geometric distributions have a few things in common with binomial distributions — but they're definitely not the same. Geometric distributions are about the question, "*How long do I have to wait to succeed once?*" But binomial distributions are about "*How often will I succeed if I have n goes?*" Geometric distributions are easier, but binomials are more common.

The Normal Distribution

Skip these two pages if you're doing OCR A S1 or OCR B S1.

The normal distribution is everywhere in statistics. Everywhere, I tell you. So learn this well...

For **Continuous** Distributions, **Area = Probability**

1) With discrete random variables, there are 'gaps' between the possible values (see page 29).
2) Continuous random variables are different — there are no gaps.
3) So for a continuous random variable, you can draw the probability density function (pdf) $f(x)$ as a line.
4) The probability of the random variable taking a value between two limits is the area under the graph between those limits.
5) This means that for any single value b, P(X = b) = 0. (Since the area under a graph at a single point is zero).
6) This also means that $P(X \le a)$ (or $P(X < a)$) is the area under the graph to the left of a. And $P(X \ge b)$ is the area to the right of b.
7) Since the total probability is 1, the total area under a pdf must also be 1.

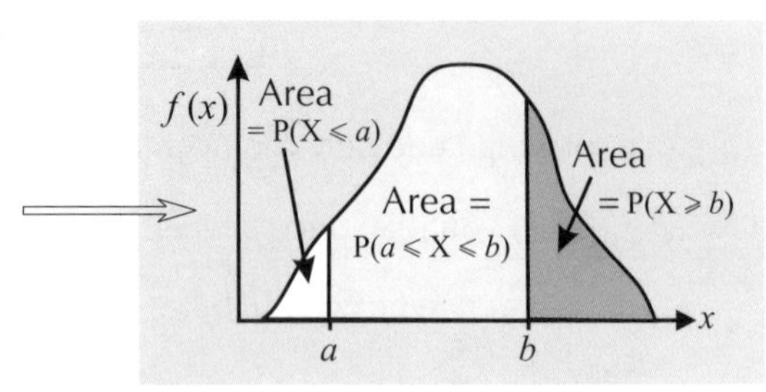

The Normal Distribution has a **Peak** in the **Middle**

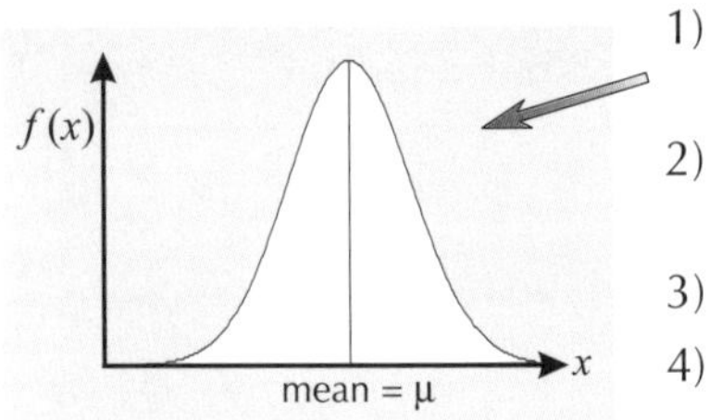

1) A random variable with a normal distribution has a pdf like this — a symmetrical bell-shaped curve.
2) The peak in the centre is at the mean (or expected value). The peak in the pdf tells you that values near the mean are most likely.
3) Further from the mean, the pdf falls — so values far from the mean are less likely.
4) The graph is symmetrical — so values the same distance above and below the mean are equally likely.
5) If X is normally distributed with **mean** μ and **variance** σ^2, it's written $X \sim N(\mu, \sigma^2)$.

Use **Normal Distribution Tables** and a **Sketch** to find probabilities

This is the practical bit.

1) Working out the area under a normal distribution curve is usually hard. But for the normally-distributed random variable Z, where $Z \sim N(0, 1)$ (Z has mean 0 and variance 1), there are tables you can use.
2) You look up a value of z and these tables (usually labelled $\Phi(z)$) tell you the probability that $Z \le z$ (which is the area under the curve to the left of z).
3) You can convert any normally-distributed variable to Z by subtracting the mean and dividing by the standard deviation — this is called normalising. This means that if you normalise a variable, you can use the Z-tables.

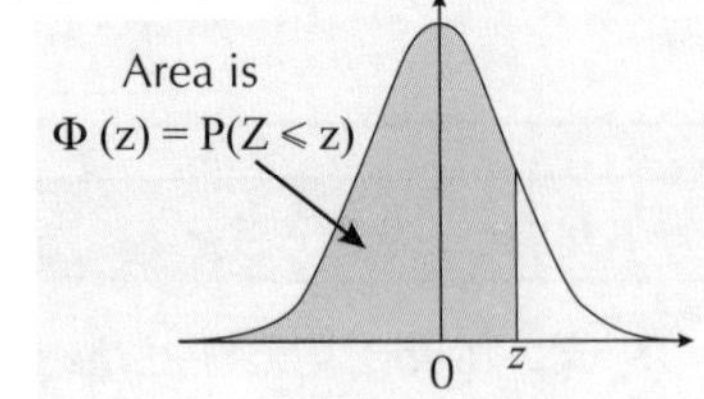

$$\text{If } X \sim N(\mu, \sigma^2)\text{, then } \frac{X-\mu}{\sigma} = Z\text{, where } Z \sim N(0, 1)$$

This means that if you subtract μ from any numbers in the question and then divide by σ — you can use your tables for Z.

Example: If $X \sim N(5, 16)$ find a) $P(X < 7)$, b) $P(X > 9)$, c) $P(5 < X < 11)$

Subtract μ (= 5) from any numbers and divide by σ (= $\sqrt{16} = 4$) — then you'll have a probability for $Z \sim N(0, 1)$.

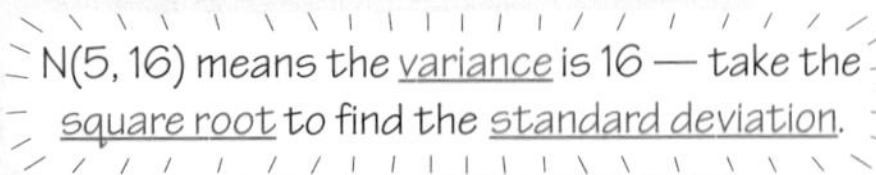

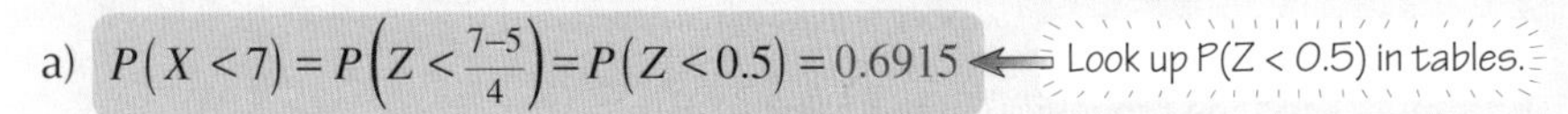

a) $P(X<7) = P\left(Z < \frac{7-5}{4}\right) = P(Z<0.5) = 0.6915$ ← Look up P(Z < 0.5) in tables.

b) $P(X>9) = P\left(Z > \frac{9-5}{4}\right) = P(Z>1) = 1 - P(Z<1) = 1 - 0.8413 = 0.1587$

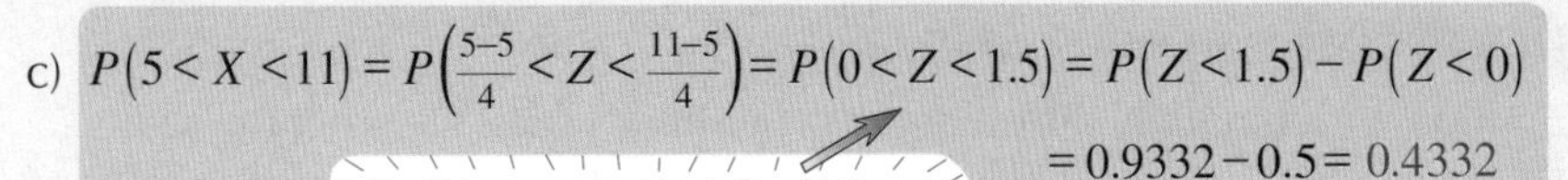

c) $P(5<X<11) = P\left(\frac{5-5}{4} < Z < \frac{11-5}{4}\right) = P(0<Z<1.5) = P(Z<1.5) - P(Z<0)$

$= 0.9332 - 0.5 = 0.4332$

Find the area to the left of 1.5 and subtract the area to the left of 0.

The Normal Distribution

You need to practise loads of questions with the kind of tables you'll be using in the Exam, since the tables from the various Exam boards all look a bit different.

Draw a Sketch with Normal Probability Questions

There's usually a 'percentage points' table to help you work backwards — use it to answer questions like: How big is k if $P(Z < k) = 0.3$?

Example: $X \sim N(53, \sigma^2)$ and $P(X < 50) = 0.2$. Find σ.

It's a normal distribution — so your first thought should be to try and normalise it.

① Subtract the mean and divide by the standard deviation:

$$P(X < 50) = P\left(Z < \frac{50-53}{\sigma}\right) = P\left(Z < -\frac{3}{\sigma}\right) = 0.2$$

Ideally, you'd look up 0.2 in the percentage points table to find $-\frac{3}{\sigma}$.
Unfortunately, in some tables it just ain't there, so you have to think a bit...

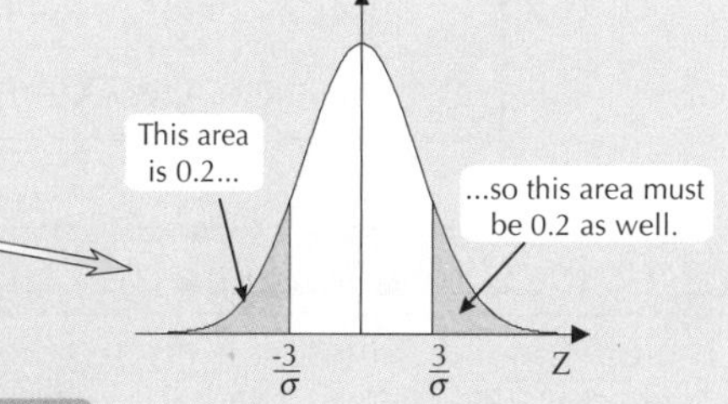

② $P\left(Z < -\frac{3}{\sigma}\right)$ is 0.2, so from the symmetry of the graph, $P\left(Z < \frac{3}{\sigma}\right)$ must be 0.8.

So look up 0.8 in the 'percentage points' table to find that an area of 0.8 is to the left of $z = 0.8416$.

This tells you that $\frac{3}{\sigma} = 0.8416$, or $\sigma = 3.56$ (to 3 sig. fig.)

Practice Questions

1) If $X \sim N(50, 16)$ find
(a) $P(X < 55)$, (b) $P(X < 42)$, (c) $P(X > 56)$ and (d) $P(47 < X < 57)$.

2) $X \sim N(600, 202)$
(a) If $P(X < a) = 0.95$, find a.
(b) If $P(|X - 600| < b) = 0.8$, find b.

Sample exam questions:

3) The exam marks of 1000 candidates are normally distributed with mean 50 marks and standard deviation 30 marks.
(a) The pass mark is 41. Estimate the number of candidates who passed the exam. [3 marks]
(b) Find the mark required for an A-grade if 10% of the candidates achieved a grade A. [3 marks]

4) The lifetimes of a particular type of battery are normally distributed with mean μ and standard deviation σ.
A student using these batteries finds that 40% last less than 20 hours and 80% last less than 30 hours.
Find μ and σ. [7 marks]

The Medium of a random variable follows a paranormal distribution...

Remember... it's definitely worth drawing a quick sketch when you're finding probabilities using a normal distribution — you're much less likely to make a daft mistake. Also, remember that it's $N(\mu, \sigma^2)$ — with the variance in the brackets and not the standard deviation. This topic isn't too bad once you're happy using the tables. So get hold of some and practise.

The Normal Distribution

Skip these two pages if you're doing OCR A S1, OCR B S1 or AQA B S1.

The normal distribution is very useful. And trust me — there will be a question on it in the Exam.

The Normal Distribution is used to Approximate the Binomial

It can be a right pain to work out numbers to do with a binomial distribution. Luckily, as long as a few conditions are met, you can use a normal distribution instead — since it'll give you pretty much the same answers.

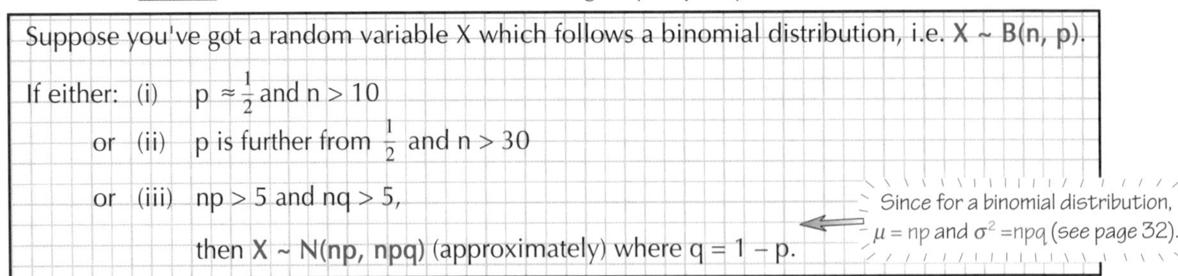

Suppose you've got a random variable X which follows a binomial distribution, i.e. **X ~ B(n, p)**.

If either: (i) $p \approx \frac{1}{2}$ and $n > 10$

or (ii) p is further from $\frac{1}{2}$ and $n > 30$

or (iii) $np > 5$ and $nq > 5$,

then **X ~ N(np, npq)** (approximately) where $q = 1 - p$.

Since for a binomial distribution, $\mu = np$ and $\sigma^2 = npq$ (see page 32).

You need to use a Continuity Correction

The binomial distribution is discrete but the normal distribution is continuous. To allow for this you need to use a continuity correction. Like a lot of this stuff, it sounds more complicated than it is.

A binomially-distributed variable X is discrete, so you can work out P(X = 0), P(X = 1), etc.

A normally-distributed variable is continuous, and so P(X = 0) = P(X = 1) = 0, etc. (see page 36).

So what you do is assume that the 'binomial 1' is spread out over the interval 0.5 - 1.5.
Then to approximate the binomial P(X = 1), you find the normal P(0.5 < X < 1.5).
Similarly, the 'binomial 2' is spread out over the interval 1.5 - 2.5 and so on.

0 ⟸1⟹⟸2⟹⟸3⟹⟸4⟹

0.5 1.5 2.5 3.5 4.5

The interval you need to use with your normal distribution depends on the binomial probability you're trying to find out.

The general principle is the same, though — each binomial value b covers the interval from $b-\frac{1}{2}$ up to $b+\frac{1}{2}$.

Binomial	Normal	
P(X = b)	$P(b - \frac{1}{2} < X < b + \frac{1}{2})$	
P(X ≤ b)	$P(X < b + \frac{1}{2})$	...to include b
P(X < b)	$P(X < b - \frac{1}{2})$	...to exclude b
P(X ≥ b)	$P(X > b - \frac{1}{2})$	...to include b
P(X > b)	$P(X > b + \frac{1}{2})$	...to exclude b

Example: If X ~ B(80, 0.4) find: (i) P(X < 45) and (ii) P(X ≥ 40)

① You need to make sure first that the normal approximation is okay to use...
$n > 30$ and p isn't too far from $\frac{1}{2}$ so the normal approximation is valid.

Or you could say that both np = 32 and nq = 48 are greater than 5.

② Next, work out np and npq: $np = 80 \times 0.4 = 32$ and $npq = 80 \times 0.4 \times (1 - 0.4) = 19.2$ ⟸ q = 1 − p
So the approximation you need is: X ~ N(32, 19.2)

③ Now you can work out your probabilities by normalising everything and using your tables for Z ($= \frac{X-32}{\sqrt{19.2}}$)

(i) You need P(X < 45) — so with the continuity correction this is P(X < 44.5).

$$P(X < 44.5) = P\left(Z < \frac{44.5-32}{\sqrt{19.2}}\right) = P(Z < 2.853) = 0.9978$$

(ii) Now you need P(X ≥ 40) — with the continuity correction this is P(X > 39.5).

$$P(X > 39.5) = P\left(Z > \frac{39.5-32}{\sqrt{19.2}}\right) = P(Z > 1.712) = 1 - P(Z < 1.712) = 1 - 0.9566 = 0.0434$$

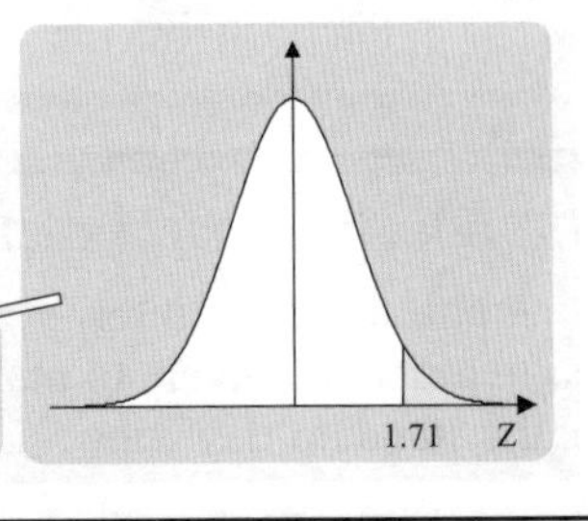

The Normal Distribution

This bit's useful. And very interesting — if you're into that kind of thing.

The Sampling Distribution of X — the Central Limit Theorem

1) Suppose you've got a random variable X with mean μ and variance σ^2.
(It's only the mean and variance you need to know — apart from this, the distribution can be anything you like.)
2) When you take a sample of n readings from the distribution of X, you can work out the sample mean $\bar{X}$.
3) If you now keep taking samples of size n from that distribution and working out the sample means, then you get a collection of sample means — these sample means will be distributed (approximately) as though they've been taken from a normal distribution with mean μ and variance $\frac{\sigma^2}{n}$. This is the **Central Limit Theorem**.
4) The bigger n is, the better this approximation will be. (For n over about 30, the approximation's pretty good.)

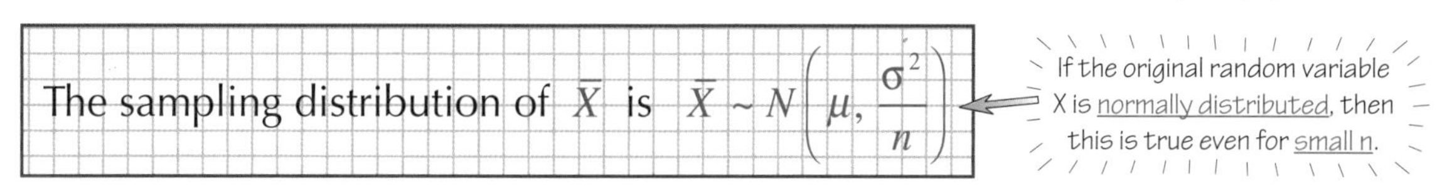

Example: A sample of size 50 is taken from a population with mean 20 and variance 10.
Find the probability that the sample mean is less than 19.

① Since n (= 50) is quite large, you can use the Central Limit Theorem. Here $\bar{X} \sim N\left(20, \frac{10}{50}\right)$ i.e. $\bar{X} \sim (20, 0.2)$

② You need $P(\bar{X} < 19)$. Since $\bar{X}$ has an (approximately) normal distribution, you can normalise it and use tables for Z. So subtract the mean from all the numbers and divide by the standard deviation (= $\sqrt{0.2}$ = 0.4472):

$$P(\bar{X} < 19) = P\left(Z < \frac{19-20}{0.4472}\right) = P(Z < -2.236)$$

③ Now it's best to draw a sketch:

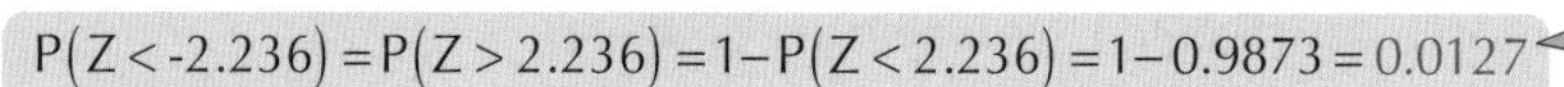

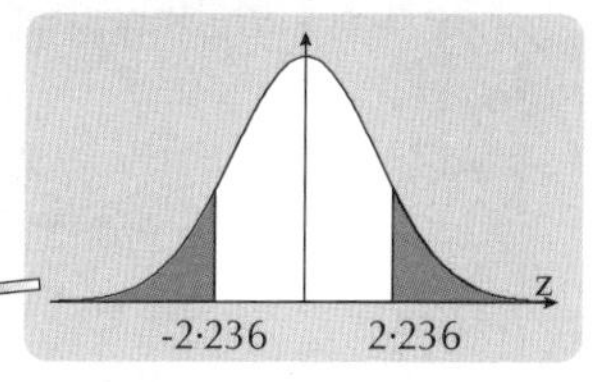

Practice Questions

1) A ~ B(200, 0.05), where A is a random variable. Use a suitable approximation to find the probability that A is less than 11.

2) 30% of the plants in a nursery are short of water. What is the probability that in a sample of 500 plants, the number of plants which are short of water is less than 149?

Sample exam questions:

3) A game involves two throws of a fair six-sided dice. If 200 people play, use a suitable approximation to find the probability that at least 10 people obtain two sixes. [6 marks]

4) A sample of 50 bags of sweets is taken from a supply which has known mean 30 g and standard deviation 5 g.
(a) State the distribution of the sample mean. [1 mark]
(b) Find the probability that the mean of the sample is less than 29 g. [2 marks]
(c) If a sample of size n bags were to be taken from the same supply, find the minimum value of n so that the probability of getting the mean of the sample to less than 29 g is less than 5%. [5 marks]

Normal, normal everywhere — not a nice thought to think...

It's all normal this and normal that at the moment. If you haven't already, then now's the time to accept that you won't get far without knowing how to answer a normal question. Personally, I can't help but think the Central Limit Theorem's pretty amazing — you can take samples from any distribution you like and say something meaningful (excuse pun) about the mean.

Correlation

Skip these two pages if you're doing OCR B S1 or AQA A S1.

Correlation is all about how closely two quantities are linked. And it can involve a fairly hefty formula.

Draw a *Scatter Diagram* to see *Patterns* in Data

Sometimes variables are measured in pairs — maybe because you want to find out how closely they're linked.
These pairs of variables might be things like: — 'my age' and 'length of my feet', or
— 'temperature' and 'number of accidents on a stretch of road'.

You can plot readings from a pair of variables on a scatter diagram — this'll tell you something about the data.

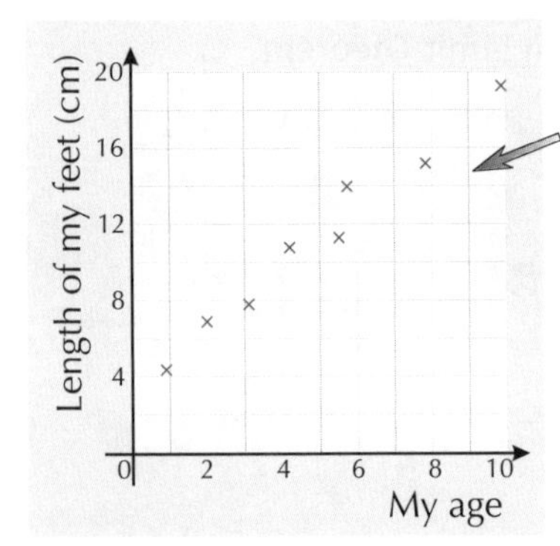

The variables 'my age' and 'length of my feet' seem linked — all the points lie close to a line. As I got older, my feet got bigger and bigger (though I stopped measuring when I was 10).

It's a lot harder to see any connection between the variables 'temperature' and 'number of accidents' — the data seems scattered pretty much everywhere.

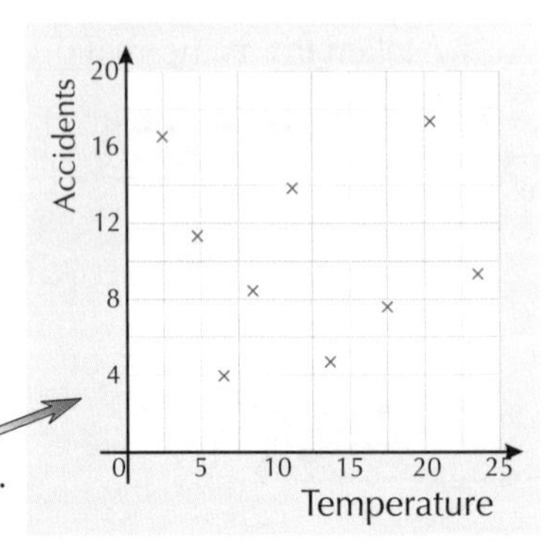

Correlation is a measure of *How Closely* variables are *Linked*

1) Sometimes, as one variable gets bigger, the other one also gets bigger — then the scatter diagram might look like the one on the right. Here, a line of best fit would have a positive gradient. The two variables are positively correlated (or there's a positive correlation between them).

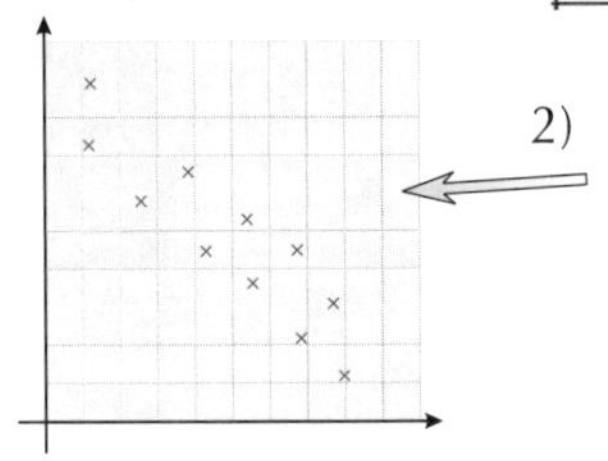

2) But if one variable gets smaller as the other one gets bigger, then the scatter diagram might look like this one — and the line of best fit would have a negative gradient. The two variables are negatively correlated (or there's a negative correlation between them).

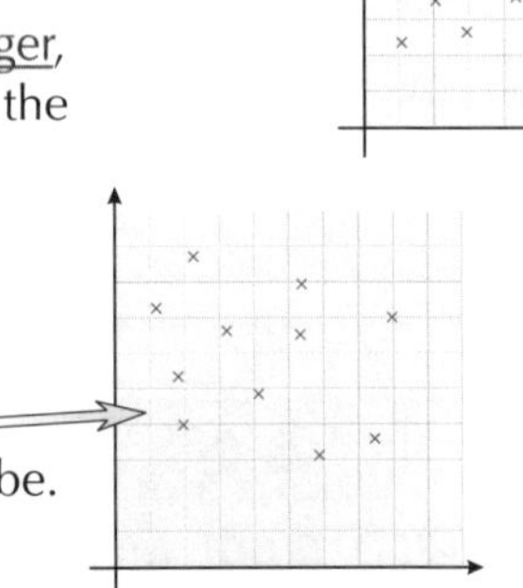

3) And if the two variables aren't linked at all, you'd expect a random scattering of points — it's hard to say where the line of best fit would be. The variables aren't correlated (or there's no correlation).

The *Product-Moment Correlation Coefficient (r)* measures Correlation

1) The Product-Moment Correlation Coefficient (PMCC, or r, for short) measures how close to a straight line the points on a scatter graph lie.

2) The PMCC is always between +1 and –1.
If all your points lie exactly on a straight line with a positive gradient (perfect positive correlation), r = +1.
If all your points lie exactly on a straight line with a negative gradient (perfect negative correlation), r = –1.
(In reality, you'd never expect to get a PMCC of +1 or –1 — your scatter graph points might lie pretty close to a straight line, but it's unlikely they'd all be on it.)

3) If r = 0 (or more likely, pretty close to 0), that would mean the variables aren't correlated.

4) The formula for the PMCC is a real stinker. But some calculators can work it out if you type in the pairs of readings, which makes life easier. Otherwise, just take it nice and slow.

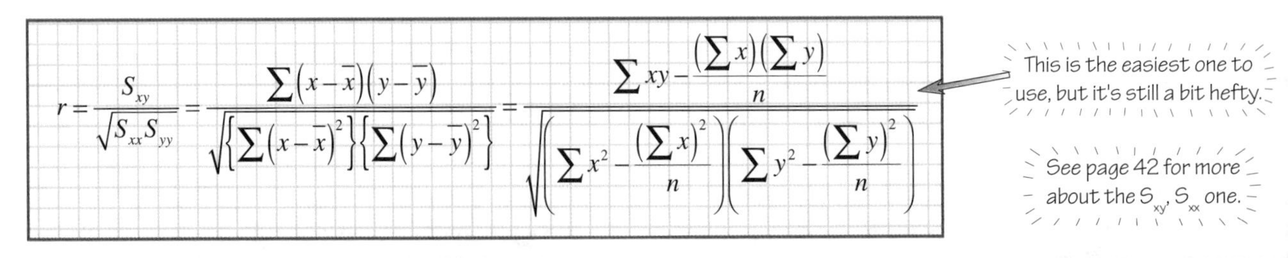

$$r = \frac{S_{xy}}{\sqrt{S_{xx}S_{yy}}} = \frac{\sum(x-\bar{x})(y-\bar{y})}{\sqrt{\left\{\sum(x-\bar{x})^2\right\}\left\{\sum(y-\bar{y})^2\right\}}} = \frac{\sum xy - \frac{(\sum x)(\sum y)}{n}}{\sqrt{\left(\sum x^2 - \frac{(\sum x)^2}{n}\right)\left(\sum y^2 - \frac{(\sum y)^2}{n}\right)}}$$

This is the easiest one to use, but it's still a bit hefty.

See page 42 for more about the S_{xy}, S_{xx} one.

Correlation

The correlation formula's a little on the large side. So don't rush this kind of question...

It's best to make a Table

Example: Illustrate the following data with a scatter diagram, and find the product-moment correlation coefficient (r) between the variables x and y.

x	1.6	2.0	2.1	2.1	2.5	2.8	2.9	3.3	3.4	3.8	4.1	4.4
y	11.4	11.8	11.5	12.2	12.5	12.0	12.9	13.4	12.8	13.4	14.2	14.3

1) The scatter diagram's the easy bit — just plot the points.

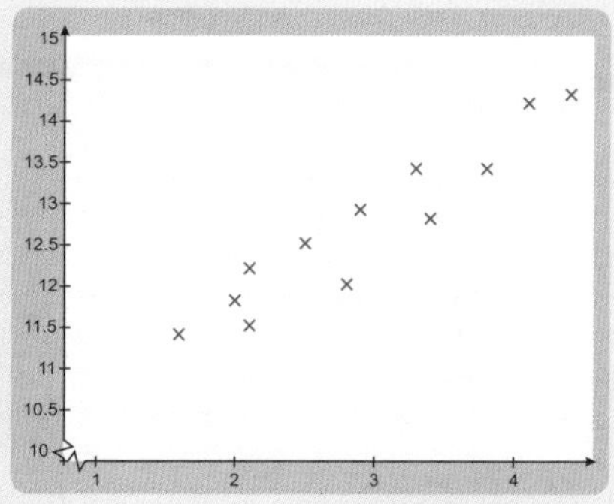

Now for the correlation coefficient. From the scatter diagram, the points lie pretty close to a straight line with a positive gradient — so if the correlation coefficient doesn't come out pretty close to +1, we'd need to worry...

2) There are 12 pairs of readings, so $n = 12$. That bit's easy — now you have to work out a load of sums. It's best to add a few extra rows to your table...

x	1.6	2	2.1	2.1	2.5	2.8	2.9	3.3	3.4	3.8	4.1	4.4	$35 = \Sigma x$
y	11.4	11.8	11.5	12.2	12.5	12	12.9	13.4	12.8	13.4	14.2	14.3	$152.4 = \Sigma y$
x^2	2.56	4	4.41	4.41	6.25	7.84	8.41	10.89	11.56	14.44	16.81	19.36	$110.94 = \Sigma x^2$
y^2	129.96	139.24	132.25	148.84	156.25	144	166.41	179.56	163.84	179.56	201.64	204.49	$1946.04 = \Sigma y^2$
xy	18.24	23.6	24.15	25.62	31.25	33.6	37.41	44.22	43.52	50.92	58.22	62.92	$453.67 = \Sigma xy$

Stick all these in the formula to get:

$$r = \frac{\left(453.67 - \frac{35 \times 152.4}{12}\right)}{\sqrt{\left(110.94 - \frac{(35)^2}{12}\right) \times \left(1946.04 - \frac{(152.4)^2}{12}\right)}} = \frac{9.17}{\sqrt{8.857 \times 10.56}} = \underline{0.948} \text{ (to 3 s.f.)}$$

This is pretty close to 1, so there's a high positive correlation between x and y.

Practice Questions

1. Plot a scatter diagram and calculate the product-moment coefficient of correlation for the data below.

Height (cm)	**165**	**176**	**159**	**167**	**174**	**171**	**169**	**168**	**169**	**172**
Weight (kg)	**72**	**90**	**70**	**75**	**86**	**84**	**80**	**81**	**82**	**83**

What does the value of the product-moment coefficient of correlation tell you about the data?

Sample Exam Question:

2. Values of two variables x and y obtained from a survey are recorded in the table below.

x	1	2	3	4	5	6	7	8
y	0.50	0.70	0.10	0.82	0.64	0.36	0.16	0.80

Represent these data on a scatter diagram, and obtain the product-moment correlation coefficient (PMCC) between the two variables.

What does this tell you about the variables? [9 marks]

What's a statistician's favourite soap — Correlation Street... (Boom boom)

It's worth remembering that the PMCC assumes that both variables are normally distributed — chances are you won't get asked a question about that, but there's always the possibility that you might, so learn it. If you know the original readings, then you can use a calculator to get the PMCC directly, but that's not an excuse for not knowing how to use the formula.

Correlation

Skip these two pages if you're doing OCR B S1 or AQA A S1.

Sometimes they don't give you all the data — just summations (things with Σ in). Just bung the numbers in the formula.

Sometimes they give you *Summary Data*

Example: I spread different amounts of manure (x) around my 10 tomato plants last summer. When I picked my toms, I measured the weight of tomatoes I got from each plant (y). Work out the correlation coefficient between x and y using the figures below.

$$\sum x = 16.5, \quad \sum y = 22.1, \quad \sum x^2 = 40.69, \quad \sum y^2 = 59.81, \quad \sum xy = 48.29$$

The correlation coefficient is sometimes written as: $r = \frac{S_{xy}}{\sqrt{S_{xx}S_{yy}}}$,

where $S_{xy} = \sum(x-\bar{x})(y-\bar{y}) = \sum xy - \frac{\sum x \sum y}{n}$ and $S_{xx} = \sum(x-\bar{x})^2 = \sum x^2 - \frac{(\sum x)^2}{n}$

S_{yy} is the same as this, but with the x's replaced with y's.

The versions in red are easier to use.

These same formulas come up on pages 44-45 as well — in regression.

Same as ever — just stick the numbers in the formula to get the correlation coefficient:

$$r = \frac{S_{xy}}{\sqrt{S_{xx}S_{yy}}} = \frac{48.29 - \frac{16.5 \times 22.1}{10}}{\sqrt{\left(40.69 - \frac{16.5^2}{10}\right)\left(59.81 - \frac{22.1^2}{10}\right)}} = \frac{11.825}{\sqrt{13.465 \times 10.969}} = \underline{0.973} \text{ (to 3 sig. fig.)}$$

The *PMCC* isn't affected by a *Linear Transformation*

Correlation coefficients aren't affected by linear transformations. In normal language, this means you can multiply variables by a number, and add a number to them — and you won't change the PMCC between them.

Example: The variables p and q are given by: $p = 4x - 3$ and $q = 9y + 17$, where x and y are as in the example above. Work out the PMCC between p and q.

Easy — p and q are linear transformations of x and y (i.e. there are no x^2 or xy bits flying around). So the PMCC between p and q is the same as the PMCC between x and y.

The PMCC between p and q is given by $\underline{r = 0.973}$ (to 3 sig. fig.)

Don't make *Sweeping Statements* using Statistics

This mini-section's all about not making 'sweeping statements' (always a temptation where statistics are concerned).

1) A high correlation coefficient doesn't necessarily mean that one quantity causes the other.

Example: The number of televisions sold in Japan and the number of cars sold in America may well be correlated, but that doesn't mean that high TV sales in Japan cause high car sales in the US.

2) The PMCC is only a measure of a linear relationship between two variables (i.e. how close they'd be to a line if you plotted a scatter diagram).

Example: In the diagram on the right, the PMCC would probably be pretty low, but the two variables definitely look linked. It looks like the points lie on a parabola (the shape of an x^2 curve) — not a line.

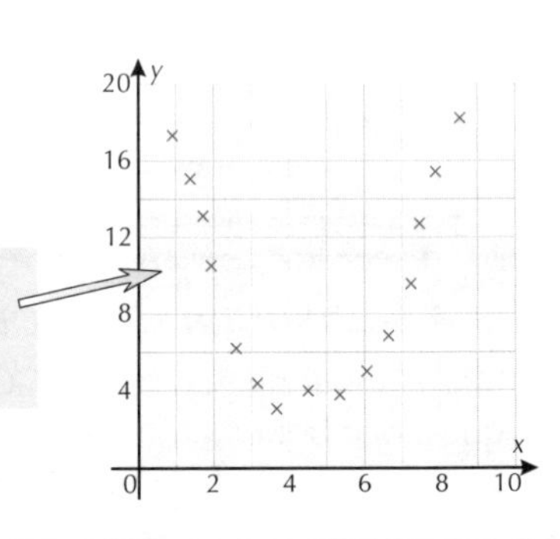

Correlation

You only need Spearman's Rank Correlation Coefficient if you're doing OCR A S1.
This is another correlation coefficient, but you can use it where you couldn't use the PMCC.

Spearman's Rank Correlation Coefficient (SRCC or r_s) works with Ranks

You can use the SRCC (or r_s, for short) when your data is a set of ranks. (Ranks are the positions of the values when you put them in order — e.g. from biggest to smallest, or from best to worst, etc.)

Example: At a dog show, two judges put 8 labradors (A-H) in the following orders, from best to worst. Calculate the SRCC between the sets of ranks.

Position	1st	2nd	3rd	4th	5th	6th	7th	8th
Judge 1:	B	C	E	A	D	F	G	H
Judge 2:	C	B	E	D	F	A	G	H

First, make a table of the ranks of the 8 labradors — i.e. for each dog, write down where it came in the show.

Dog	A	B	C	D	E	F	G	H
Rank from Judge 1:	4	1	2	5	3	6	7	8
Rank from Judge 2:	6	2	1	4	3	5	7	8

Now for each dog, work out the difference (d) between the ranks from the two judges — you can ignore minus signs.

Dog	A	B	C	D	E	F	G	H
d	2	1	1	1	0	1	0	0

Take a deep breath, and add another line to your table — this time for d^2:

Dog	A	B	C	D	E	F	G	H	Total = Σd^2
d^2	4	1	1	1	0	1	0	0	8

Then the SRCC is:

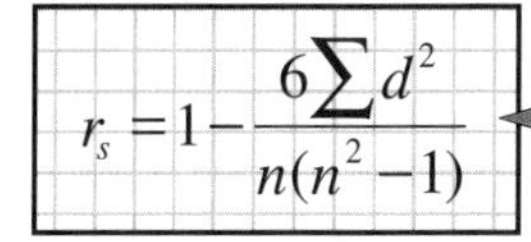

$$r_s = 1 - \frac{6\sum d^2}{n(n^2-1)}$$

You can ignore minus signs when you work out d, since only d^2 is used to work out the SRCC.

So here, $r_s = 1 - \frac{6\times 8}{8\times(8^2-1)} = 1 - \frac{48}{504} = 0.905$ (to 3 sig. fig.)

— this is close to +1, so the judges ranked the dogs in a pretty similar way.

If there'd been 2 or more equal values, you'd need to find the average rank. E.g. if Judge 1 had awarded joint 7th place to G and H, then they'd each have rank $\frac{7+8}{2} = 7.5$ (i.e. they would have been 7th and 8th)

Practice Questions

1. ***These are the marks obtained by 10 pupils in their Physics and English exams. Calculate Spearman's rank correlation coefficient.***

Physics	54	34	23	58	52	58	13	65	69	52
English	16	73	89	81	23	81	56	62	61	37

Sample Exam question:

2. A chocolate manufacturer sells eight products. The sales of each product in a certain period (y) and the amount of money spent advertising each product (x) were recorded.
   (a) Using the information below, calculate the product-moment correlation coefficient between the variables 'sales' and 'advertising costs'.
   $\sum x = 386,\ \sum y = 460,\ \sum x^2 = 25426,\ \sum y^2 = 28867,\ \sum xy = 26161$ [5 marks]
   (b) What does this result tell you about the quantities x and y? [2 marks]

I don't like this page — it's a bit rank...

The good thing about the SRCC is that you can use it even when your data isn't normally distributed. More generally, if you're anything like me, then you'll find it easy to scribble a load of important-looking numbers down on a piece of paper, and then forget what any of them actually are. So do yourself a favour — take tricky questions slowly, be nice and methodical, and write down every step of your working clearly. It makes it easier to find any mistakes later.

Linear Regression

Skip these two pages if you're doing OCR B S1 or AQA A S1.

Linear regression is just fancy stats-speak for 'finding lines of best fit'. Not so scary now, eh...

Decide which is the *Independent Variable* and which is the *Dependent*

Example: The data below show the load on a lorry, x (in tonnes), and the fuel consumption, y (in km per litre).

x	5.1	5.6	5.9	6.3	6.8	7.4	7.8	8.5	9.1	9.8
y	9.6	9.5	8.6	8.0	7.8	6.8	6.7	6	5.4	5.4

1) The variable along the x-axis is the explanatory or independent variable — it's the variable you can control, or the one that you think is affecting the other.
   The variable 'load' goes along the x-axis here.

2) The variable up the y-axis is the response or dependent variable — it's the variable you think is being affected.
   In this example, this is the fuel consumption.

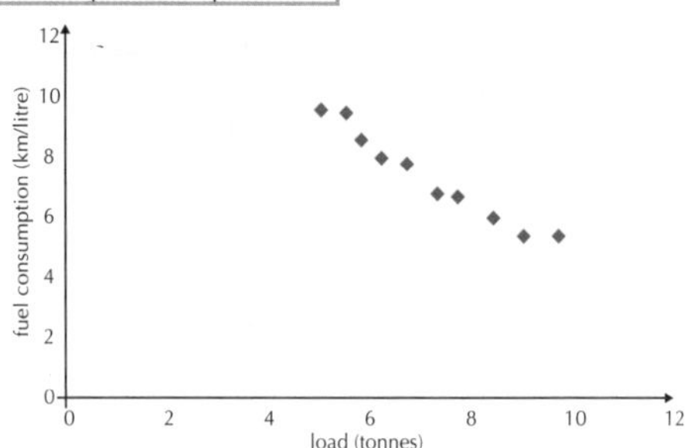

The *Regression Line* (Line of Best Fit) is in the form *y = a + bx*

To find the line of best fit for the above data you need to work out some sums.
Then it's quite easy to work out the equation of the line. If your line of best fit is $y = a + bx$, this is what you do...

① First work out these four sums — a table is probably the best way: $\sum x,\ \sum y,\ \sum x^2,\ \sum xy$.

x	5.1	5.6	5.9	6.3	6.8	7.4	7.8	8.5	9.1	9.8	$72.3 = \Sigma x$
y	9.6	9.5	8.6	8	7.8	6.8	6.7	6	5.4	5.4	$73.8 = \Sigma y$
x^2	26.01	31.36	34.81	39.69	46.24	54.76	60.84	72.25	82.81	96.04	$544.81 = \Sigma x^2$
xy	48.96	53.2	50.74	50.4	53.04	50.32	52.26	51	49.14	52.92	$511.98 = \Sigma xy$

② Then work out S_{xy}, given by: $S_{xy} = \sum(x-\bar{x})(y-\bar{y}) = \sum xy - \dfrac{(\sum x)(\sum y)}{n}$

and S_{xx}, given by: $S_{xx} = \sum(x-\bar{x})^2 = \sum x^2 - \dfrac{(\sum x)^2}{n}$

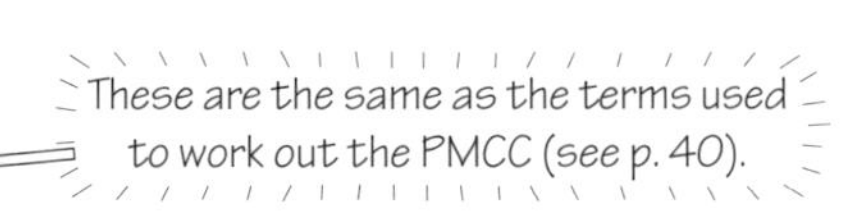

③ The gradient (b) of your regression line is given by: $b = \dfrac{S_{xy}}{S_{xx}}$

④ And the intercept (a) is given by: $a = \bar{y} - b\bar{x}$.

⑤ Then the regression line is just: $y = a + bx$.

Loads of calculators will work out regression lines for you — but you still need to know this method, since they might give you just the sums from Step 1.

Example: Find the equation of the regression line of y on x for the data above.

The 'regression line of y on x' means that x is the independent variable, and y is the dependent variable.

1) Work out the sums: $\sum x = 72.3,\ \sum y = 73.8,\ \sum x^2 = 544.81,\ \sum xy = 511.98$.

2) Then work out S_{xy} and S_{xx}: $S_{xy} = 511.98 - \dfrac{72.3 \times 73.8}{10} = -21.594 \qquad S_{xx} = 544.81 - \dfrac{72.3^2}{10} = 22.081$

3) So the gradient of the regression line is: $b = \dfrac{-21.594}{22.081} = -0.978$ (to 3 sig. fig.)

Remember: $\bar{x} = \dfrac{\sum x}{n}$

4) And the intercept is: $a = \dfrac{\sum y}{n} - b\dfrac{\sum x}{n} = \dfrac{73.8}{10} - (-0.978) \times \dfrac{72.3}{10} = 14.451$

The regression line always goes through the point $(\bar{x}, \bar{y})$.

5) This all means that your regression line is: $y = 14.451 - 0.978x$

Linear Regression

Residuals — the difference between Practice and Theory

A residual is the difference between an observed y-value and the y-value predicted by the regression line.

Residual = Observed y-value – Estimated y-value

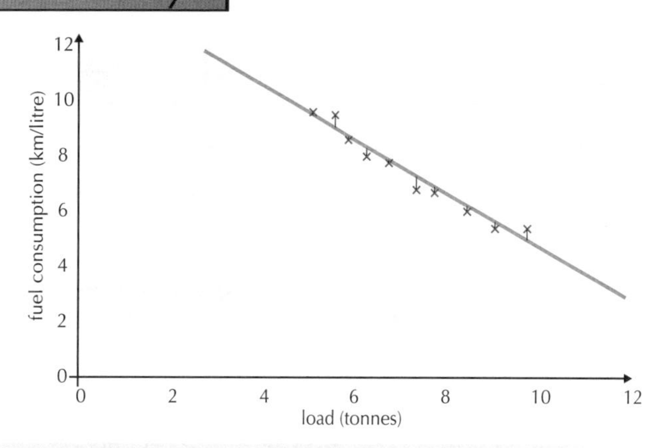

1) Residuals show the experimental error between the y-value that's observed and the y-value your regression line says it should be.
2) Residuals are shown by a vertical line from the actual point to the regression line.

Example: For the fuel consumption example opposite, calculate the residuals for: (i) x = 5.6, (ii) x = 7.4.

(i) When x = 5.6, the residual = 9.5 – (-0.978 × 5.6 + 14.451) = 0.526 (to 3 sig. fig.)
(ii) When x = 7.4, the residual = 6.8 – (-0.978 × 7.4 + 14.451) = -0.414 (to 3 sig. fig.)

A positive residual means the regression line is too low for that value of x.
A negative residual means the regression line is too high.

This kind of regression line is called Least Squares Regression, because you're finding the equation of the line which minimises the sum of the squares of the residuals (i.e. $\sum e_k^2$ is as small as possible, where the e_k are the residuals).

Use Regression Lines With Care

You can use your regression line to predict values of y. But it's best not to do this for x-values outside the range of your original table of values.

Example: Use your regression equation to estimate the value of y when: (i) x = 7.6, (ii) x = 12.6

(i) When x = 7.6, $y = -0.978 \times 7.6 + 14.451 = 7.02$ (to 3 sig. fig.). This should be a pretty reliable guess, since x = 7.6 falls in the range of x we already have readings for — this is called interpolation.
(ii) When x = 12.6, $y = -0.978 \times 12.6 + 14.451 = 2.13$ (to 3 sig. fig.). This may well be unreliable since x = 12.6 is bigger than the biggest x-value we already have — this is called extrapolation.

Practice Questions

Sample Exam question:

1. The following times (in seconds) were taken by eight different runners to complete distances of 20 metres and 60 metres.

Runner	A	B	C	D	E	F	G	H
20-metre time (x)	3.39	3.20	3.09	3.32	3.33	3.27	3.44	3.08
60-metre time (y)	8.78	7.73	8.28	8.25	8.91	8.59	8.90	8.05

(a) Plot a scatter diagram to represent the data. [3 marks]
(b) Find the equation of the regression line of y on x and plot it on your scatter diagram. [8 marks]
(c) Use the equation of the regression line to estimate the value of y when: (i) $x = 3.15$, (ii) $x = 3.88$ and comment on the reliability of your estimates. [6 marks]
(d) Find the residuals for: (i) $x = 3.32$, (ii) $x = 3.27$. Illustrate them on your scatter diagram. [4 marks]

99% of all statisticians make sweeping statements...

Be careful with that extrapolation business — it's like me saying that because I grew at an average rate of 10 cm a year for the first few years of my life, by the time I'm 50 I should be 5 metres tall. Residuals are always errors in the values of y — these equations for working out the regression line all assume that you can measure x perfectly all the time.

Null and Alternative Hypotheses

You only need to learn these pages if you're doing OCR B S1.

This section's all about hypothesis testing. It doesn't sound much fun, but it's dead useful.
And you need to concentrate — this isn't the most straightforward stuff in the book...

Hypothesis Tests check whether *Real Data* supports your *Theory*

The basic idea behind hypothesis testing isn't so bad — it's all the details that make things tricky.

1) Hypothesis testing involves testing a hypothesis (theory) you might have about a situation.
2) In a nutshell, you say what your hypothesis is, collect some data, and then see if the data backs up your theory.

State your *Null Hypothesis* and *Alternative Hypothesis* at the start

1) Whenever you carry out a hypothesis test, you need to be very clear about the theory you're testing. The first thing you have to do is state a null hypothesis (called H_0).
2) You have to be specific about your null hypothesis because you initially assume that it's true, and do all your calculations based on it. So the null hypothesis is always an equality.

   You assume the null hypothesis is true until you have reason to doubt it.

3) For example, you might want to test whether a coin is 'fair' when you toss it. In this case, you'd be doing a test connected with a binomial distribution (e.g. the number of heads in 10 tosses). Your null hypothesis could be that the probability of a head (p) equals 0.5 (which is what you'd expect if the coin is fair).
4) But as well as having a null hypothesis, you have to state an alternative hypothesis (or H_1). If you need to reject the null hypothesis (i.e. if the data you collect doesn't back it up), you have to have an alternative ready.

You can have *Different Kinds* of *Alternative Hypothesis*

Your null hypothesis is often pretty straightforward, but you generally have a choice of alternative hypotheses.

1) The number of heads you get when you toss a coin will have a binomial distribution (see page 32). So if you toss it 10 times, the number of heads will be described by X, where $X \sim B(10, p)$.
2) A hypothesis test to see if the coin is fair is a test about the value of p. If the coin is fair, then $p = 0.5$. (This is your null hypothesis, i.e. H_0: $p = 0.5$.)
3) Then you have a choice of alternative hypotheses...

   Maybe you just want to know whether the coin is fair.
   If the coin isn't fair, then $p \neq 0.5$ — and your alternative hypothesis would be: $H_1: p \neq 0.5$

   You're not really bothered whether $p > 0.5$ or $p < 0.5$.

   But maybe you want to know whether the coin is biased towards heads. If it's biased towards heads, then $p > 0.5$. In this case, your alternative hypothesis is: $H_1: p > 0.5$

   The alternative hypothesis is usually an inequality — involving $\neq, <, >, \geq$ or $\leq$.

   Or you might want to know whether the coin is biased towards tails.
   If this is the case, then $p < 0.5$. Then your alternative hypothesis would be: $H_1: p < 0.5$

Different *Alternative* hypotheses lead to *One-Tail* and *Two-Tail* tests

Two-Tail tests have a ≠ sign in H_1:

If H_1 has a '≠' sign, then you can reject H_0 for two reasons — either your experimental results are too big to believe H_0, or they're too small. This is called a two-tail test.

E.g. $H_0: p = 0.5 \quad H_1: p \neq 0.5$

One-Tail tests have one of >, <, ≥, ≤ in H_1:

If H_1 has one of $>, <, \geq$ or $\leq$, then you have a one-tail test.
If H_1 has a $>$ or $\geq$ sign, then you'd reject H_0 if your results are too big to believe H_0 (i.e. the alternative hypothesis is better for big results).
But if H_1 has a $<$ or $\leq$ sign, then you can only reject H_0 because your experimental results are too small.

E.g. $H_0: p = 0.5 \quad H_1: p > 0.5$

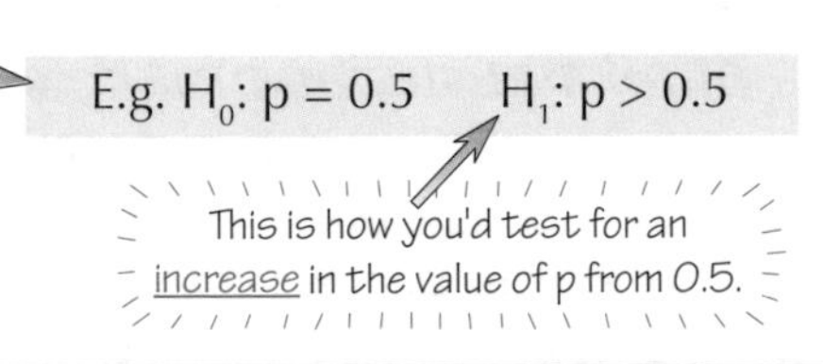

You'd use a one-tail test when you're testing for either an increase or a decrease in the value of p.

Null and Alternative Hypotheses

Get your head round these examples of null and alternative hypotheses — they're all tests about p, where p is the probability of success in the binomial B(n, p). And make sure you understand why they are one-tail or two-tail tests.

You can do different tests with a **Binomial Distribution**

Example 1: Test whether the value of p for a binomial distribution equals 0.4.

You're testing whether p equals 0.4 — so the null hypothesis is easy:

$H_0: p = 0.4$

The alternative is that it doesn't equal 0.4 — so your alternative hypothesis is:

$H_1: p \neq 0.4$

This is a two-tail test — you can reject H_0 if p is either too big or too small.

Example 2: Test whether the value of p has increased from 0.2.

Remember, the null hypothesis has to be an equality — so the only possibility for H_0 is:

$H_0: p = 0.2$

If this null hypothesis is true, then p has stayed the same. But you want to reject this H_0 if p has increased, so your alternative hypothesis would be:

$H_1: p > 0.2$

This is a one-tail test — you can only reject H_0 if p is too big.

Example 3: Test whether the value of p is not less than 0.7.

Again, the null hypothesis has to be an equality — so H_0 is:

$H_0: p = 0.7$

This time, you want to reject H_0 if there's evidence that p is less than 0.7. So your H_1 is:

$H_1: p < 0.7$

This is a one-tail test — you can only reject H_0 if p is too small.

Practice Questions

1. ***Write down null and alternative hypotheses that you'd use to carry out the tests on p described below. State in each case whether your hypotheses lead to a one-tail or a two-tail test.***
   - ***a)*** ***p is 0.35.***
   - ***b)*** ***p is at least 0.5. (So you want to reject H_0 if p is greater than 0.5.)***
   - ***c)*** ***p is at most 0.1.***
   - ***d)*** ***p has increased from 0.4. (So you want to reject your null hypothesis of 'no change' if p has increased.)***
   - ***e)*** ***p has fallen from 0.2.***

Useful quotes: I never put off till tomorrow what I can do the day after*...

Getting the null hypothesis right (the equality) is usually a piece of cake. The alternative hypothesis can be tricky, though. It's sometimes easier to think about what you want to conclude if you reject H_0 — then use that as H_1. So if you want to show that p has increased from 0.8 (so $H_0: p = 0.8$), use $H_1: p > 0.8$ — then rejecting H_0 means you conclude p is now bigger.

* *Oscar Wilde*

Critical Region and Significance Level

You only need to learn these pages if you're doing OCR B S1.

So far, so good, eh. Well keep going — it doesn't really get any worse.

The **Critical Region** contains the **Results** that would make you **Reject H_0**

1) If you're doing a hypothesis test, you assume your null hypothesis is true until you have reason to doubt it.
2) So basically you collect some data, and ask, "*How likely is this data if the null hypothesis is true?*"
3) If it's really unlikely that you'd get those results under H_0 (i.e. if H_0 is true), then maybe H_0 isn't true after all.

Example: (The same old '*Is the coin fair?*' hypothesis test.) I'm going to collect data by tossing the coin 10 times. So if X is the number of heads, $X \sim B(10, p)$. My hypotheses are:

$$H_0: p = 0.5 \qquad H_1: p \neq 0.5$$

i) If the null hypothesis is true, then the probability of getting 5 heads is $\binom{10}{5} 0.5^5 (1-0.5)^5 = 0.246$.

That's quite high — so basically I wouldn't be at all surprised if I got 5 heads.

ii) And if the null hypothesis is true, then the probability of getting 1 head is $\binom{10}{1} 0.5^1 (1-0.5)^9 = 0.00977$.

That's not impossible, but it's very low. So if H_0 is true, I'd be very surprised if I only got 1 head. In fact, I'd be so surprised, I probably wouldn't believe that my null hypothesis was actually true. I'd be more inclined to believe my alternative hypothesis, i.e. that $p \neq 0.5$.

4) This is the idea behind a critical region. The critical region is essentially a set of results that would be so unlikely under your null hypothesis that you'd start to believe your alternative hypothesis is more likely to be true — i.e. you'd reject your null hypothesis.
5) If you're doing a two-tail test, then your critical region is in two parts. One part contains the results that are too low to believe that H_0 is true — the other part is the set of results that are too high.

The **Significance Level** of a test is ***P(reject $H_0|H_0$ true)***

1) In the example above, getting 1 head under the null hypothesis wasn't impossible — just very unlikely.
2) The significance level of a hypothesis test is a measure of how unlikely a result has to be before you reject (i.e. disbelieve) your null hypothesis.
3) Very often, a significance level of 0.05 (or 5%) is used. This means that if the probability of a result occurring under your null hypothesis is less than 0.05, then you'd reject H_0 and go with H_1, the alternative hypothesis.
4) You can think of the significance level in another way. Suppose your null hypothesis really is true. If you have a significance level of 0.05, then the probability of getting a result in your critical region (and so rejecting H_0) is 0.05. This is unlikely, but not impossible. So the significance level is Prob(reject $H_0|H_0$ true) — the probability of rejecting H_0 given that H_0 is true.
5) The most important thing to remember when you're looking for your critical region is that you need the results whose probability is less than your significance level.
6) Also, you have to be careful with two-tail tests, since your critical region is in two halves. If your significance level is 5%, then the probability that you reject H_0 because your experimental results are too high is 2.5% (0.025). Similarly, the probability that you reject H_0 because your results are too low is 2.5%.

Critical region (2.5%)
Acceptance region
Critical region (2.5%)

The **Test Statistic** is a **Result** from an **Experiment**

1) The test statistic is just a value that occurs in an experiment that lets you test your null hypothesis.
2) In the 'fair coin' example, the test statistic would be 'number of heads'.

Hypothesis Testing

Remember the 4 steps of a Hypothesis Test

Step 1 — State your null hypothesis (H_0) and alternative hypothesis (H_1).
Step 2 — Decide on your significance level.
Step 3 — Determine the test statistic, and find the critical region for that significance level.
Step 4 — Get your data. And if your test statistic falls in the critical region, reject the null hypothesis.

Right then... here we go — and this time it's for real... the '*Is the coin fair?*' test in all its glory.

Example: Describe a two-tail hypothesis test to determine whether a coin is fair. Use a significance level of 5%.

1. The null hypothesis is easy — assume the coin is fair until you find out otherwise.

H_0: p = 0.5

It's a two-tail test, so the alternative hypothesis is: H_1: p ≠ 0.5

This means the critical region is in two parts. You'll reject H_0 if the test statistic is either too high or too low.

2. The question says to use a significance level of 5% (or 0.05).

3. The test statistic is the number of heads.
Since this is a two-tail test, the critical region is in two halves — one half is the set of results that are too high to believe that H_0 is true, and the other half is the set of results that are too low.

Note all the red inequality signs below — you always need to find probabilities that X is less than or greater than certain 'cut-off points' when you're finding critical regions.

This is half the total significance level, since it's a two-tail test.

i) To find the bottom half of the critical region, you need a value for k so that $P(X < k) < 0.025$.
It's best to use binomial tables for this. Since you're tossing the coin 10 times, use the n = 10 table.
From tables: $P(X \leq 2 \mid p = 0.5) = 0.055$ — this is greater than 0.025, so a result of '2 heads' wouldn't be 'surprising enough' to actually reject H_0.
But $P(X \leq 1 \mid p = 0.5) = 0.011$ — this is less than 0.025, so the lower critical region is '1 or fewer heads'.

ii) You find the upper part of the critical region in the same way, so you need q so that $P(X \geq q) < 0.025$.
From tables: $P(X \geq 8 \mid p = 0.5) = 0.055$ — this is greater than 0.025, so 8 isn't in the critical region.
But: $P(X \geq 9 \mid p = 0.5) = 0.011$ — this is less than 0.025, so 9 is in the critical region.
So the upper critical region is '9 or more heads'.

iii) So the overall critical region will be '1 or fewer heads' and '9 or more heads'.

Example: For this 'fair coin' test, what would be the conclusion if you threw the coin 10 times and got:
a) 9 heads? b) 3 heads?

a) 9 heads is inside the critical region, so you'd reject H_0 — the evidence suggests that p ≠ 0.5 (i.e. that the alternative hypothesis is more likely to be true). In other words, the evidence suggests that the coin is unfair.

b) 3 heads is outside the critical region, so you wouldn't reject H_0 — the results aren't unlikely enough to make you disbelieve H_0. In other words, you don't have sufficient evidence to say that the coin is unfair.

Practice Questions

1. Find the critical regions for the following hypothesis tests on p, where X ~ B(n, p). Use a significance level of 5% in all cases. The data for the test will be collected from 20 trials.
a) H_0: p = 0.4, H_1: p < 0.4. b) H_0: p = 0.3, H_1: p > 0.3.
c) H_0: p = 0.2, H_1: p ≠ 0.2. (Hint: Find the lower critical region and the upper critical region separately.)

H_0: I think — hence failure to reject H_0 ⇒ I am...

You can also think of the significance level as the 'standard of proof' you need before you reject H_0. If you can reject H_0 using a significance level of 5%, then that's pretty good evidence that H_0 isn't true. However, if you can reject H_0 using a significance level of 1%, then that's much stronger evidence that H_0 is untrue. There's more about this on the next page.

Hypothesis Testing Using B(n, p)

You only need to learn these pages if you're doing OCR B S1.

Hypothesis testing isn't the easiest thing in S1. But I reckon it's probably the most interesting bit.

Hypothesis Tests are used in ***Real Life***

Use the good ol' 4-step method from page 49 whenever you're asked to do one of these hypothesis tests.

Example:

When items being produced in a factory were checked by quality control, it was found that 25% of the items produced were faulty. This was unacceptable and so the machinery was overhauled. After this overhaul a random sample of 20 items was taken from the production line and only 2 of them were found to be faulty. The manager claimed that the overhaul had therefore been effective, and had reduced the percentage of faulty items being produced.

Carry out a hypothesis test at the 5% significance level to test the manager's claim that the value of p has fallen.

It's a big question, so you need to take your time and remember the four steps...

(1)

1) Let the 'number of faulty items' be given by X, where $X \sim B(n, p)$. Then this hypothesis test is a test about the value of p.
2) The first thing to do is state the null and alternative hypotheses.
3) Originally, 25% of the items were faulty. So if p is the probability that each item is faulty, then originally $p = 0.25$.
4) The claim is that p has fallen. So the thing to do here is assume that it's still the same, and see if you can disprove that. This means you need the null hypothesis:

   $H_0: p = 0.25$ ← Remember, H_0 must be an equality.

5) You want to reject H_0 ('no change') if p has fallen. So your alternative hypothesis needs to be:

   $H_1: p < 0.25$ ← You're testing for a reduction in the value of p.

(2) This is a one-tail test with a significance level of 5% (0.05).

(3)

1) The test statistic is the 'number of faulty items'. Since a sample of 20 items was used, this will be distributed as B(20, 0.25) — assuming H_0 is true.
2) This is a one-tail test, so the critical region is in one piece — and consists of all the results less than or equal to *k*, where: $P(X \leq k) < 0.05$.
3) Using tables: P(2 or fewer faulty) = $P(X \leq 2) = 0.091$ — this is greater than the significance level, and so is not in the critical region.

   P(1 or fewer faulty) = $P(X \leq 1) = 0.024$ — this is less than the significance level, and so is in the critical region.
4) So the critical region for this test is '*1 or fewer items faulty*'.

(4)

1) When a random sample of items was taken, 2 were found to be faulty.
2) This is not in the critical region, so you can't reject H_0.
3) This means that there isn't enough evidence to reject H_0 at this level of significance. And since H_0 was the hypothesis that p was the same as before, the evidence doesn't support the manager's claim that the overhaul was effective.

1) Note that if you used a significance level of 10%, your conclusion would be different — you'd reject the null hypothesis of 'no change' and you'd say the evidence supported the manager's claim.
2) This is because when you use a significance level of 10%, you're asking for 'less proof' of the need to reject H_0.
3) It depends on whether you want 'some evidence' or 'strong evidence' before you reject H_0.

Hypothesis Testing Using B(n, p)

Example:

When 800 seeds of a type of plant were sown in compost, 240 of them germinated.

(i) Write down an estimate of p, the probability a seed of this type will germinate in compost.

A firm claims that if these seeds are planted in a new type of soil they sell, then you can expect more seeds to germinate than if they were sown in compost.

(ii) Design a hypothesis test at the 5% significance level to test this claim. To collect the results, 20 seeds will be sown in the new type of soil, and the number germinating counted.

Draw a diagram, in the form of a number line, showing the critical region.

(iii) State the conditions necessary for this test to be valid.

(i) $p = P(\text{seed germinates}) = \frac{240}{800} = 0.3$

(ii) And now for the hypothesis test...

① $H_0 : p = 0.3$
$H_1 : p > 0.3$

You want to reject the null hypothesis of 'no difference' if enough seeds germinate.

② The significance level is 5% (0.05), and it's a one-tail test.

③ The test statistic will be 'number of seeds that germinate'.

The critical region is numbers greater than or equal to k, where $P(X \geq k) < 0.05$, and $X \sim B(20, 0.3)$.

From tables: P(9 or more germinate) = 1 – P(8 or fewer) = 1 – 0.8867 = 0.1133 > 0.05
— this is greater than the significance level, so 9 isn't in the critical region.

But: P(10 or more germinate) = 1 – P(9 or fewer) = 1 – 0.9520 = 0.0480 < 0.05
— this is less than the significance level, so 10 is in the critical region.

So you'd reject the null hypothesis (of 'no difference') if 10 or more seeds germinate.

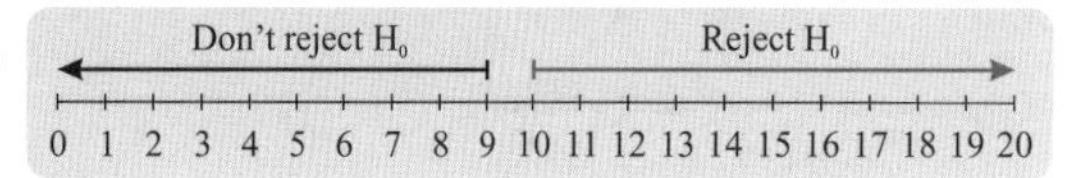

(iii) For this test to be valid the 20 seeds sown should be a random sample from all the seeds available to choose from, the probability of each seed germinating should be the same, and whether a particular seed germinates should be independent of what happens to any other seed.

Practice Questions

1. A random sample of 500 people was selected, and each person was asked which party they intended to vote for at the next election. The *People's Party* was the choice of 300 people.

(a) Write down an estimate for p, the probability that a person intends to vote for the *People's Party*. [1 mark]

After a party political broadcast by the *People's Party*, a survey of 20 people chosen at random was carried out. It was found that 17 of the sample intended to vote for the *People's Party*. The campaign manager claimed that the broadcast had been a success and that the value of p had been increased.

You are to carry out a hypothesis test at the 5% significance level to test this claim.

(b) State the hypotheses used to test this claim. [2 marks]

(c) Find the critical region for this test. Explain your reasoning. [3 marks]

(d) Is the campaign manager's claim supported by the survey? State your conclusion carefully. [2 marks]

Useful quotes: Either that wallpaper goes or I do...*

Hypothesis-testing questions can look scary at first glimpse, but if you use the 4-step method, you won't go far wrong. So that's it then... another module over. And what a module it was... full of ups, downs, heartbreak and sorrow. But there were good times too. No, I'm sure there were... they came near the start of the book, but they were there. Honest...

* Last words of Oscar Wilde

Answers

Section One — Data

Page 3

1) 12.8, 13.2, 13.5, 14.3, 14.3, 14.6, 14.8, 15.2, 15.9, 16.1, 16.1, 16.2, 16.3, 17.0, 17.2 (all in cm)

2)

Boys		Girls
9, 5	1	2
7, 6, 2, 0	2	1, 4, 5, 7, 8, 9
9, 7, 4, 2	3	1, 6, 7, 8, 9

Key 2|1|3 means: Boys 12, Girls 13

3)

Length of call	Lower class boundary (lcb)	Upper class boundary (ucb)	Class width	Frequency	Frequency density = Height of column
0 - 2	0	2.5	2.5	10	4
3 - 5	2.5	5.5	3	6	2
6 - 8	5.5	8.5	3	3	1
9 - 15	8.5	15.5	7	1	0.143

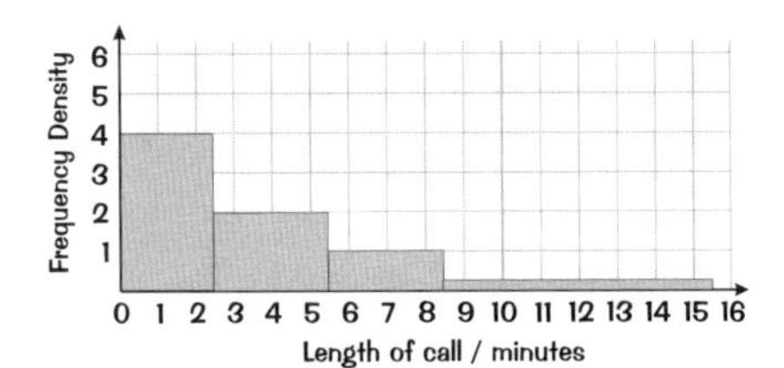

4) a)

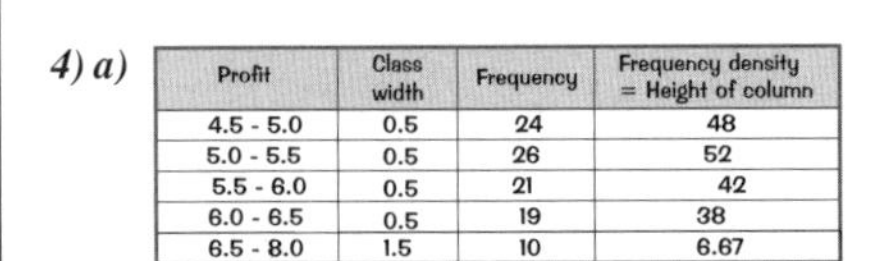

Profit	Class width	Frequency	Frequency density = Height of column
4.5 - 5.0	0.5	24	48
5.0 - 5.5	0.5	26	52
5.5 - 6.0	0.5	21	42
6.0 - 6.5	0.5	19	38
6.5 - 8.0	1.5	10	6.67

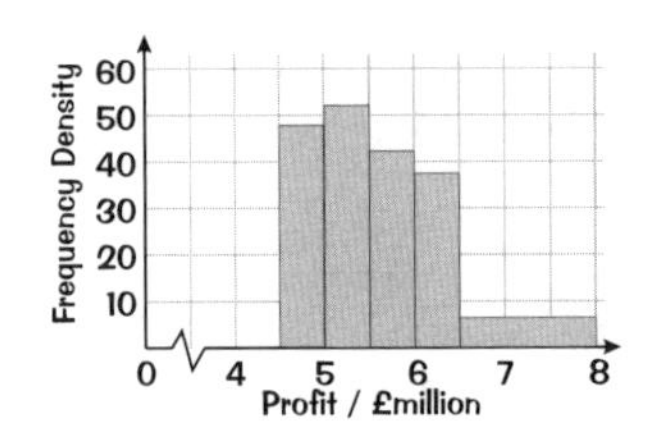

[1 mark for correct axes, then up to 2 marks for the bars drawn correctly]

b) *The distribution is positively skewed — only a few businesses make a high profit. The modal profit is between £5 million and £5.5 million.*
[Up to 2 marks available for any sensible comments]

Page 5

1) *$\Sigma f = 16$, $\Sigma fx = 22$, so mean = 22 ÷ 16 = 1.375*
Median position = 17 ÷ 2 = 8.5, so median = 1
Mode = 0.

2)

Speed	mid-class value x	Number of cars f	fx
30 - 34	32	12 (12)	384
35 - 39	37	37 (49)	1369
40 - 44	42	9	378
45 - 50	47.5	2	95
	Totals	60	2226

Estimated mean = 2226 ÷ 60 = 37.1 mph
Median position is 61 ÷ 2 = 30.5.
This is in class 35 - 39.
30.5 – 12 = 18.5, so median is 18.5th value in class.
Class width = 5, so median is:

$$34.5 + \left(\frac{18.5}{37} \times 5\right) = 37 \text{ mph}$$

Modal class is 35 - 39 mph.

3) a) *There are 30 males, so median is in 31 ÷ 2 = 15.5th position. Take the mean of the 15th and 16th readings to get median = (62 + 65) ÷ 2 = 63.5 [1 mark]*

b) *The female median is 64.5 (halfway between the 8th and 9th readings). Female median is higher than the male median. The females scored better than the males on average.*
Female range = 79 – 55 = 24.
Male range = 79 – 43 = 36
The female range is less than the male range. Their scores are more consistent than the males'.
[Up to 2 marks available for any sensible comments]

Page 7

1)

Distance	Upper class boundary (ucb)	f	Cumulative frequency (cf)
	0	0	0
0 - 2	2	10	10
2 - 4	4	5	15
4 - 6	6	3	18
6 - 8	8	2	20

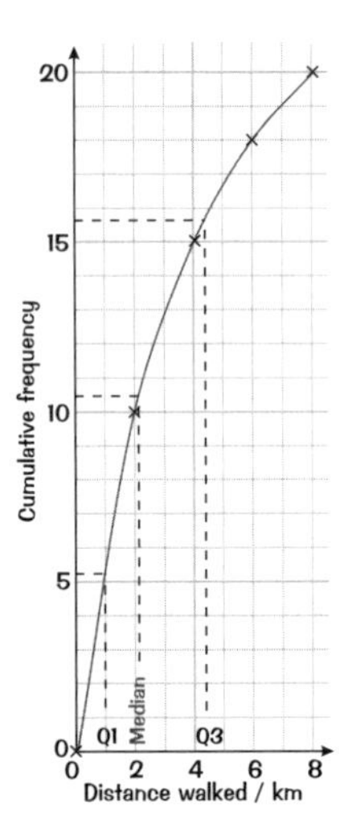

Median = 2.2 km (approximately)
Q_1 = 1 km, Q_3 = 4.4 km, so interquartile range = 3.4 km (approximately)

2) a)

Age	Upper class boundary (ucb)	f	Cumulative frequency (cf)
Under 5	5	0	0
5 - 10	11	2	2
11 - 15	16	3	5
16 - 20	21	10	15
21 - 30	31	2	17
31 - 40	41	2	19
41 - 70	71	1	20

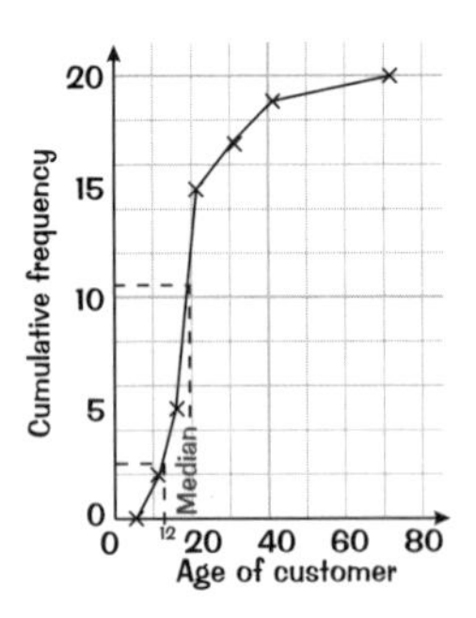

[1 mark for correctly labelled axes, up to 2 marks for calculating the points and plotting them correctly, and 1 mark for joining the points with a suitable line]

b) (i) *Median = 18.5 years (approximately) [1 mark]*
(ii) *Number of customers under 12 = 2 (approximately) [1 mark]*
So number of customers over 12 = 20 – 2 = 18 (approximately) [1 mark]

3) a) (i) *Times = 2, 3, 4, 4, 5, 5, 5, 7, 10, 12*
Median position = 5.5, so median = 5 minutes [1 mark]
(ii) *Lower quartile = 4 minutes [1 mark]*
Upper quartile = 7 minutes [1 mark]

b)

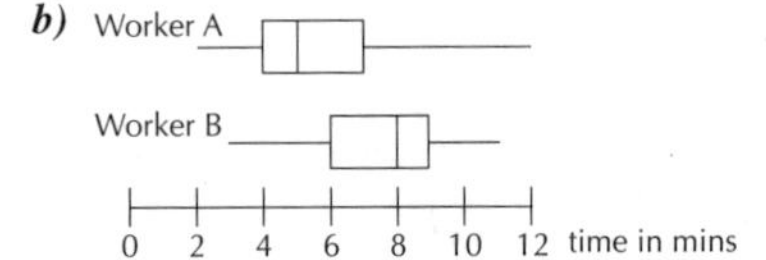

[Up to 3 marks available for each graph — get 1 mark for the median in the right place, 1 mark for both of the quartiles shown correctly, and 1 mark for the lines showing the extremes drawn correctly]

Answers

c) *Various statements could be made, e.g. the times for Worker B are longer than those for Worker A, on average. The IQR for both workers is the same — generally they both work with the same consistency. The range for Worker A is larger than that for Worker B. Worker A has a few items he/she can iron very quickly and a few which take a long time. [1 mark for any sensible answer]*

d) *Worker A would be best to employ. The median time is less than for Worker B, and the Upper Quartile is less than the median of Worker B. Worker A would generally iron more items in a given time than worker B. [1 mark for any sensible answer]*

Page 9

1) $Mean = \frac{11+12+14+17+21+23+27}{7} = \frac{125}{7} = 17.9$ *to 3 sig. fig.*

$$s.d. = \sqrt{\frac{11^2+12^2+14^2+17^2+21^2+23^2+27^2}{7} - \left(\frac{125}{7}\right)^2} = \sqrt{30.98} = 5.57 \text{ to 3 sig. fig.}$$

2)

Score	Mid-class value, x	x^2	f	fx	fx^2
100 - 106	103	10609	6	618	63654
107 - 113	110	12100	11	1210	133100
114 - 120	117	13689	22	2574	301158
121 - 127	124	15376	9	1116	138384
128 - 134	131	17161	2	262	34322
		Totals	50 (= n)	5780 (= Σx)	670618 (= Σx^2)

$$Mean = \frac{5780}{50} = 115.6$$

$$s^2 = \frac{670618}{50} - 115.6^2 = 49$$

So $s = 7$

3) a) $\bar{A} = \frac{60.3}{20} = 3.015\ g$ *[1 mark]*

b) $s_A^2 = \frac{219}{20} - 3.015^2 = 1.86\ g^2$

So $s_A = 1.36\ g$ *to 3 sig. fig.*

[3 marks for the correct answer — otherwise 1 mark for a correct method to find the variance, and 1 mark for taking the square root to find the s.d.]

c) *Brand A chocolate drops are heavier on average than brand B. Brand B chocolate drops are much closer to the mean of 2.95 g. [1 mark for each of 2 sensible statements]*

d) $Mean\ of\ A\ and\ B = \frac{\sum A + \sum B}{50} = \frac{60.3 + (30 \times 2.95)}{50} = 2.976\ g$

$$\frac{\sum B^2}{30} - 2.95^2 = 1, \text{ and so } \sum B^2 = 291.075$$

$$Variance\ of\ A\ and\ B = \frac{\sum A^2 + \sum B^2}{50} - 2.976^2 = \frac{219 + 291.075}{50} - 2.976^2$$

$= 1.3449$

So $s.d. = \sqrt{1.3449} = 1.16\ g$ *to 3 sig. fig*

[4 marks for the correct answer — otherwise 1 mark for the correct total mean, 1 mark for the sum of the B^2*, 1 mark for the correct method to find the total variance/standard deviation]*

Page 11

1) *Let* $y = x - 20$.

Then $\bar{y} = \bar{x} - 20$ *or* $\bar{x} = \bar{y} + 20$

$\sum y = 125$ *and* $\sum y^2 = 221$

So $\bar{y} = \frac{125}{100} = 1.25$ *and* $\bar{x} = 1.25 + 20 = \underline{21.25}$

$s_y^2 = \frac{221}{100} - 1.25^2 = 0.6475$ *and so* $s_y = 0.805$ *to 3 sig. fig.*

Therefore $s_x = 0.805$ *to 3 sig. fig.*

2)

Time	Mid-class x	$y = x - 35.5$	f	fy	fy^2
30 - 33	31.5	-4	3	-12	48
34 - 37	35.5	0	6	0	0
38 - 41	39.5	4	7	28	112
42 - 45	43.5	8	4	32	256
		Totals	20 (= n)	48 (= Σy)	416 (= Σy^2)

$$\bar{y} = \frac{48}{20} = 2.4$$

So $\bar{x} = \bar{y} + 35.5 = 2.4 + 35.5 = \underline{37.9 \text{ minutes}}$

$s_y^2 = \frac{416}{20} - 2.4^2 = 15.04$*, and so* $s_y = 3.88$ *minutes, to 3 sig. fig.*

But $s_x = s_y$*, and so* $\underline{s_x = 3.88 \text{ minutes, to 3 sig. fig.}}$

3) a) *Let* $y = x - 30$.

$\bar{y} = \frac{228}{19} = 12$ *and so* $\underline{\bar{x} = \bar{y} + 30 = 42}$

$s_y^2 = \frac{3040}{19} - 12^2 = 16$ *and so* $s_y = 4$

But $s_x = s_y$ *and so* $\underline{s_x = 4}$

[3 marks for the correct answers — otherwise 1 mark for the correct mean and 1 mark for the correct s.d. or variance of the y]

b) $\bar{x} = \frac{\sum x}{19} = 42$

And so $\sum x = 42 \times 19 = \underline{798}$

$$s_x^2 = \frac{\sum x^2}{19} - \bar{x}^2 = \frac{\sum x^2}{19} - 42^2 = 16$$

And so $\sum x^2 = (16 + 42^2) \times 19 = \underline{33820}$

[3 marks for both correct answers — otherwise 1 mark for either correct]

c) *New* $\sum x = 798 + 32 = 830$.

So new $\bar{x} = \frac{830}{20} = \underline{41.5}$

New $\sum x^2 = 33820 + 32^2 = 34844$.

So new $s_x^2 = \frac{34844}{20} - 41.5^2 = 19.95$ *and new* $\underline{s_x = 4.47 \text{ to 3 sig. fig.}}$

[2 marks for each correct answer — otherwise 1 mark for some correct working for each part]

Page 13

1) *IQR = 88 – 62 = 26, so 3 × IQR = 78.*
So upper fence = 88 + 78 = 166.
This means that: ***a)*** *161 is not an outlier.*
b) *176 is an outlier.*
Lower fence = 62 – 78 = -16.
This means that: ***c)*** *0 is not an outlier.*

2) *Put the 20 items of data in order:*
1, 4, 5, 5, 5, 5, 5, 6, 6, 7, 7, 8, 10, 10, 12, 15, 20, 20, 30, 50
Then the median position is 10.5, and since the 10th and the 11th items are both 7, the median = *7.*
Lower quartile = *5.*
Upper quartile = (12 + 15) ÷ 2 = *13.5.*

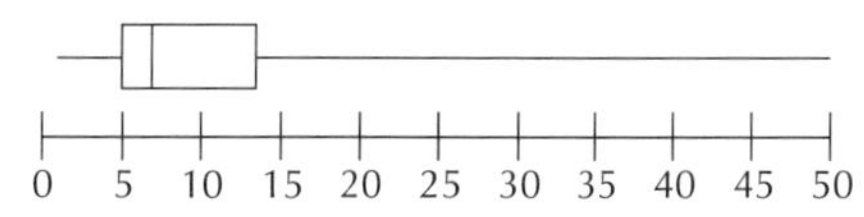

This data is positively skewed. Most 15-year-olds earned a small amount of pocket money. A few got very large amounts.

Answers

3) *Pearson's coefficient of skewness* $= \frac{10.3 - 10}{1.5} = 0.2$

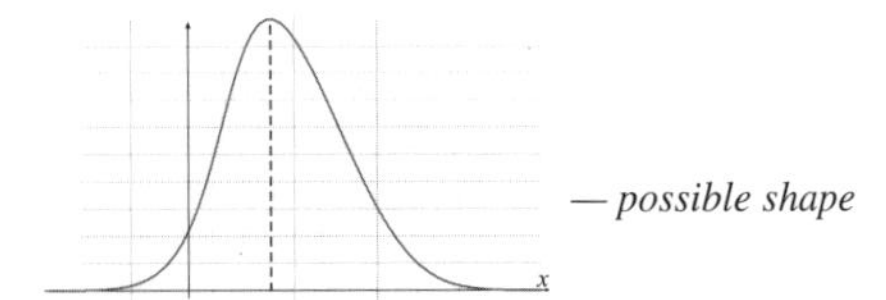

— possible shape

This is positively skewed, but not by much.

4) a) *Total number of people = 38.*
Median position = (38 + 1) ÷ 2 = 19.5.
19th value = 15; 20th value = 16, so median = 15.5 hits.
Mode = 15 hits.
[2 marks for the correct median, otherwise 1 mark for some correct working, plus 1 mark for the correct mode]

b) *Lower quartile = 10th value = 14*
Upper quartile = 29th value = 17. [1 mark for both]
So interquartile range = 17 – 14 = 3,
and upper fence = 17 + (3 × 3) = 26.
This means that 25 is not an outlier. [1 mark]

c)

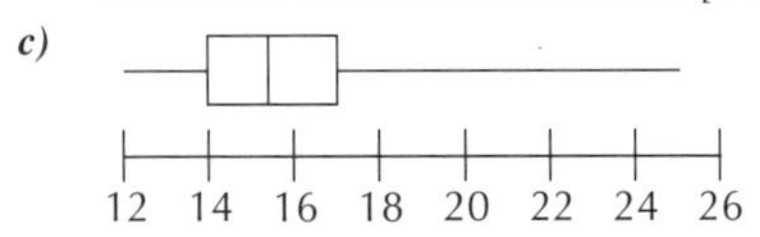

[1 mark]
The distribution is positively skewed. [1 mark]
(Different kinds of sketch would be allowed.)

d) *If 25 was removed then the right-hand tail of the box plot would be much shorter, and the distribution would be less positively skewed. [1 mark]*

5) a)

mm of rain	Upper class boundary (ucb)	f	Cumulative frequency (cf)
Under 5	5	0	0
5 - 10	10	2	2
10 - 15	15	3	5
15 - 20	20	5	10
20 - 25	25	7	17
25 - 30	30	10	27
30 - 35	35	3	30

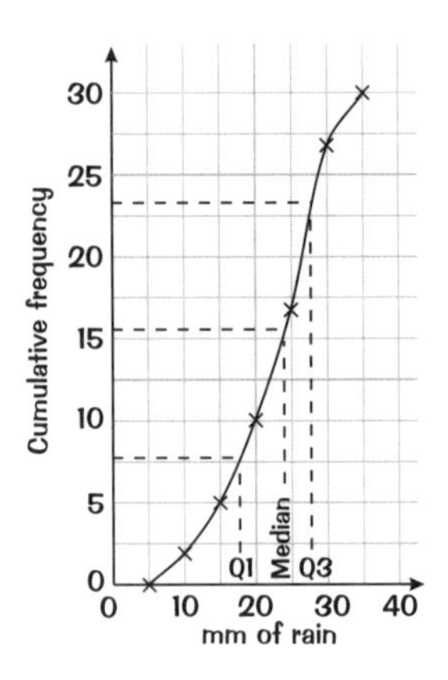

[Up to 2 marks for correctly calculating and plotting the points, plus 1 mark for correct axes/labels etc.]

b) *From the diagram, median = 24. [1 mark for answer in range 23.5-24.5]*
Lower quartile = 17.5. [1 mark for answer in the range 17-18]
Upper quartile = 27.5. [1 mark for answer in the range 27-28]

c) *Quartile coefficient of skewness* $= \frac{27.5 - (2 \times 24) + 17.5}{27.5 - 17.5} = -0.3$

[1 mark — answers may differ slightly, depending on answers to part b)]
The graph is negatively skewed — most of the days tend to have higher rainfall. [1 mark]

Page 15

1) *First do 100 ÷ 20 = 5.*
Pick a start number between 1 and 5 (by rolling a dice, for example, but roll again if you get a 6).
Then take every 5th value after that.
So if you start at the second value, your sample would be the 2nd, 7th, 12th, 17th values, etc.

2) a) *The population is everybody who attended the party (500 students).*

b) *Obtaining information from all 500 students.*

c) *An advantage is that you will get more accurate results. A disadvantage is that it is time-consuming, and it may not be easy to get a response from everyone.*

3) *One suggestion would be to...*
Split the population into:
age groups (e.g. 18-25, 25-40, over 40)
and gender groups (male and female).
Then you might try and get 20 responses from 18-25 year old females, 20 responses from 18-25 year old males, 20 from 25-40 year old females, 20 from 25-40 year old males, etc.
In total there are 6 groups, and so you'd have 120 responses in total.
The interviewer would have to stop people in the street in a busy shopping centre (for example) until they have filled their quota.

4) a) *Mean* $= \frac{Total}{n} = \frac{347}{25} = 13.88$ *cm [1 mark]*

b) *25 ÷ 5 = 5.*
First number the items of data (so call the first row of data items 1-5, the second row items 6-10, etc.)
Then start somewhere between 1 and 5 — e.g. roll a dice and get a 3 (say).
This means you include the 3rd item, 12, in your sample.
Then take every 5th value, so you'd also include 17, 14, 12 and 13.
So the full sample would be 12, 17, 14, 12 and 13.
(Or you could just select a random row/column.)
[Up to 2 marks for a correct answer]

Page 17

1) *Number each item with a 3-digit number from 000 to 999.*
Then use a random number table (or the Ran# button on a calculator) to generate 100 different numbers. If a number is repeated, discount it and choose another.

2) *Roll a dice to obtain a starting point — 5 say. Then take a 3-digit number (using 000 to represent 1000). In this case, the winning ticket would be number 839.*

3) a) *Total number of visits = 86.*

Population mean $= \frac{86}{25} = 3.44$. *[1 mark]*

b) *For example...*
I numbered the top row of numbers 1-15, and the second row 16-25.
Then I used a random number generator to get the following list of numbers between 0 and 25:
3, 9, 23, 13, 7, 7, 13, 24, 16, 19, 2, 21...
(I had to disregard the second occurrences of 7 and 13.)
Then I took the 3rd item, the 9th item, the 23rd item etc., and included them in my sample.
So my sample was: 4, 9, 0, 2, 1, 0, 1, 0, 12, 0.
[Up to 2 marks for a suitable answer]

c) *The population is divided (stratified) into 2 very different groups (car owners and those without a car), and this sample might not represent them in the correct proportions. [1 mark]*

d) (i) *Number of car owners* $= \frac{15}{25} \times 10 = 6$

Number of non car owners $= \frac{10}{25} \times 10 = 4$

[1 mark for each]
(ii) *Number the car owners 1-15. Use a random number generator to select 6 numbers between 1 and 15 — I got: 15, 9, 10, 8, 5 and 7.*
So my sample included the following data items: 4, 9, 10, 8, 4 and 1.
Then number the non car owners 1-10. Use a random number generator to select 4 numbers between 1 and 10 — I got: 6, 5, 9 and 4.
So my sample included the following data items: 0, 2, 0 and 0.
[Up to 2 marks for a suitable answer]

e) *Mean of simple random sample* $= \frac{29}{10} = 2.9$ *[1 mark]*

Mean of stratified random sample $= \frac{38}{10} = 3.8$ *[1 mark]*

Answers

The mean of the stratified random sample is much bigger. The simple random sample contained too many people who don't own a car. (Your answer may be different — it all depends on your samples.) [1 mark]

Section Two — Probability

Page 19

1) a) *The sample space would be as below:*

		Dice					
		1	2	3	4	5	6
Coin	H	2	4	6	8	10	12
	T	5	6	7	8	9	10

b) *There are 12 outcomes in total, and 9 of these are more than 5, so P(score >5) = 9/12 = 3/4*

c) *There are 6 outcomes which have a tail showing, and 3 of these are even, so P(even score given that you throw a tail) = 3/6 = 1/2*

2) a) *20% of the people eat chips, and 10% of these is 2% — so 2% eat both chips and sausages.*
Now you can draw the Venn diagram:

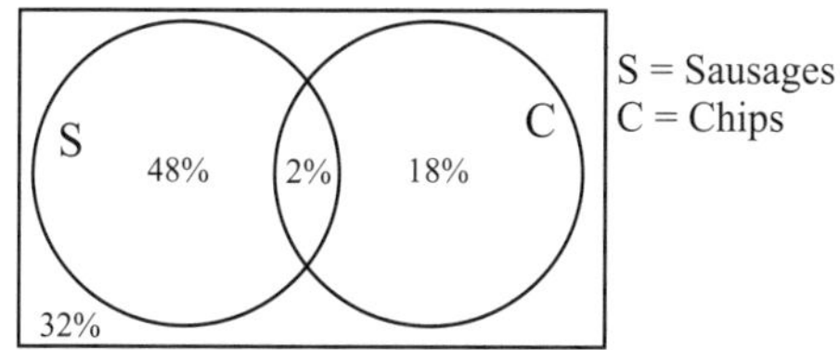

By reading the numbers in the appropriate sets from the diagram you can see

b) *18% eat chips but not sausages.*

c) *18% + 48% = 66% eat chips or sausages, but not both.*

3) a) *The Venn diagram would look something like this:*

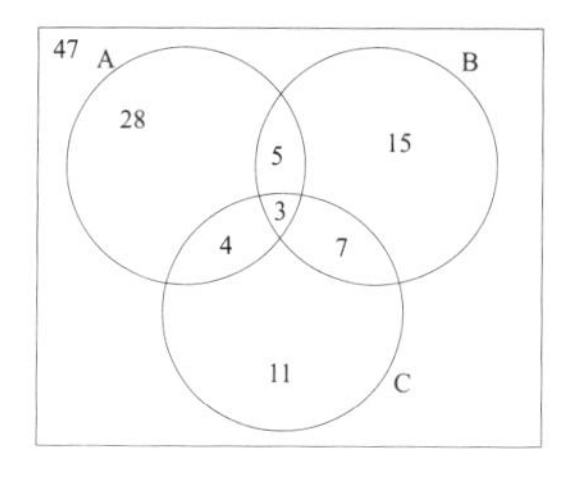

[1 mark for the central figure correct, 2 marks for '5', '7' and '4' correct (get 1 mark for 2 correct), plus 2 marks for '28', '15' and '11' correct (get 1 mark for 2 correct)]

b) *Add up the numbers in all the circles to get 73 people out of 120 buy at least 1 type of soap. So the probability = 73/120 [2 marks]*

c) *Add up the numbers in the intersections to get 5 + 3 + 4 + 7 = 19, meaning that 19 people buy at least two soaps, so the probability a person buys at least two types = 19/120 [2 marks]*

d) *28 + 11 + 15 = 54 people buy only 1 soap, and of these 15 buy soap B [1 mark]. So probability of a person who only buys one type of soap buying type B is 15/54 = 5/18 [2 marks]*

Page 21

1) *Draw a sample space diagram*

2nd Dice						
6	7	8	9	10	11	12
5	6	7	8	9	10	11
4	5	6	7	8	9	10
3	4	5	6	7	8	9
2	3	4	5	6	7	8
1	2	3	4	5	6	7
	1	2	3	4	5	6
	1st Dice					

There are 36 outcomes altogether.

a) *15 outcomes are prime (since 2, 3, 5, 7 and 11 are prime), so P(prime) = 15/36 = 5/12*

b) *7 outcomes are square numbers (4 and 9), so P(square) = 7/36*

c) *Being prime and a square number are exclusive events, so P(prime or square) = 15/36 + 7/36 = 22/36 = 11/18*

2) a)

B (0.4): U (0.3), L (0.7)
G (0.6): U (0.5), L (0.5)

b) *Choosing an upper school pupil means either 'boy and upper' or 'girl and upper'. P(boy and upper) = 0.4 × 0.3 = 0.12.*
P(girl and upper) = 0.6 × 0.5 = 0.30.
So P(Upper) = 0.12 + 0.30 = 0.42.

3) a)

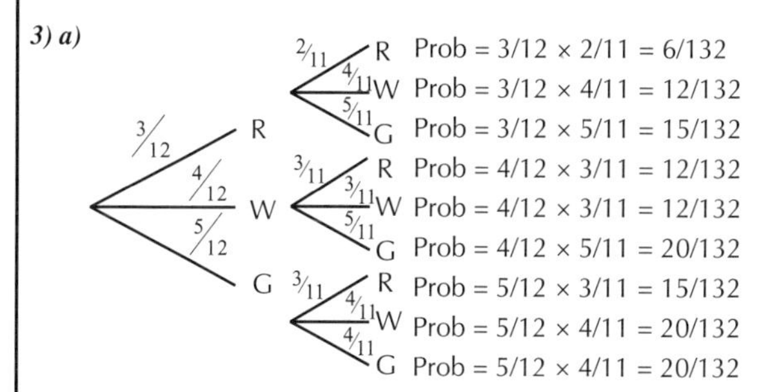

[4 marks available — 1 mark for each set of branches correct]

b) *The second counter is green means one of three outcomes 'red then green' **or** 'white then green' **or** 'green then green'. So Prob(2nd is green) = 15/132 + 20/32 + 20/132 = 55/132 = 5/12 [2 marks]*

c) *For both to be red there's only one outcome: 'red then red' Prob(both red) = 6/132 = 1/22 [2 marks]*

d) *'Both same colour' is the complementary event of 'not both same colour'.*
So Prob (not same colour) = 1 – P(both same colour)
Both same colour is either R and R or W and W or G and G
Prob(not same colour) = 1 – [6/132 + 12/132 + 20/132]
= 1 – 38/132 =94/132 = 47/66
[3 marks for the correct answer — otherwise up to 2 marks available for using a suitable method]

Page 23

1) *You could draw a Venn diagram — but you don't have to, it just makes things easier.*

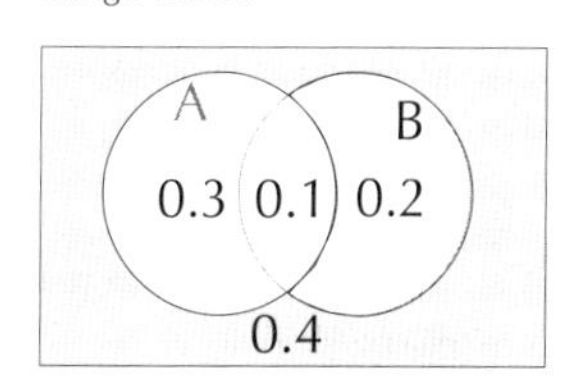

a) *P(B|A) = P(A∩B)÷ P(A) = 0.1÷0.4 = 0.25*

b) *P(B|A') = P(A'∩B)÷P(A') = 0.2÷0.6 = 1/3*

c) *P(B|A) = 0.25, but P(B) = 0.3 so they're not independent.*

2) *Draw a tree diagram:*

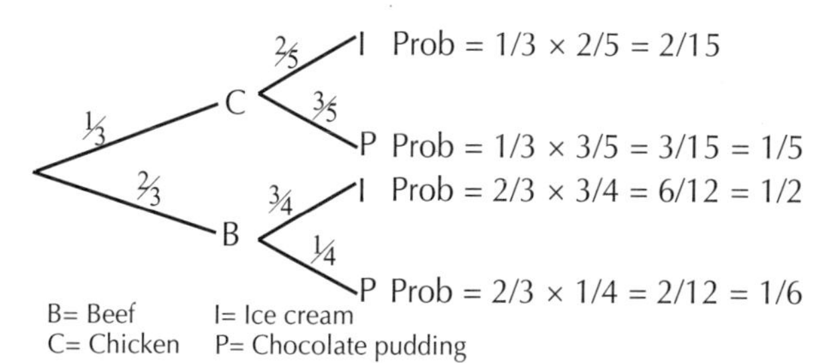

a) *P(chicken or ice cream but not both) = P(C∩P) + P(B∩I) = 1/5 + 1/2 = 7/10*

b) *P(ice cream) = P(C∩I) + P(B∩I) = 2/15 + 1/2 = 19/30*

c) *P(chicken|ice cream) = P(C∩I) ÷ P(I)= (2/15) ÷ (19/30) = 4/19*

Answers

3) a) **(i)** *V and W are independent, so*
$P(V \cap W) = P(V) \times P(W) = 0.2 \times 0.6 = 0.12$ *[1 mark]*
(ii) $P(V \cup W) = P(V) + P(W) - P(V \cap W) = 0.2 + 0.6 - 0.12 = 0.68$
[2 marks]

b) $P(U|V') = P(U \cap V') \div P(V')$
Now $U = V' \cap W'$, *so* $U \cap V' = U$ *— think about it — all of U is contained in V', so* $U \cap V'$ *(the 'bits in both U and V') are just the bits in U.*
Therefore $P(U \cap V') = P(U) = 1 - P(V \cup W) = 1 - 0.68 = 0.32$
And so $P(U|V') = 0.32 \div 0.8 = 0.4$
[3 marks for the correct answer — otherwise up to 2 marks available for correct working]

4) *Draw a tree diagram*

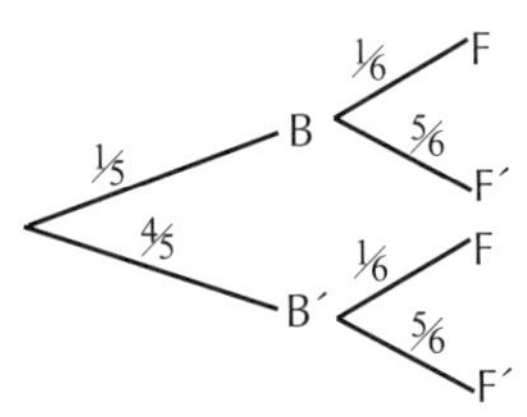

B= Biased dice shows 6
F= Fair dice shows 6

a) $P(B') = 0.8$ *[1 mark]*

b) *Either at least one of the dice shows a 6 or neither of them do, so these are complementary events. Call F the event 'the fair dice shows a 6'.*
Then $P(F \cup B) = 1 - P(F' \cap B') = 1 - (4/5 \times 5/6) = 1 - 2/3 = 1/3$
[2 marks]

c) $P(\text{exactly one 6} \mid \text{at least one 6}) = P(\text{exactly one 6} \cap \text{at least one 6}) \div P(\text{at least one 6})$.
This next step might be a bit easier to get your head round if you draw a Venn diagram.

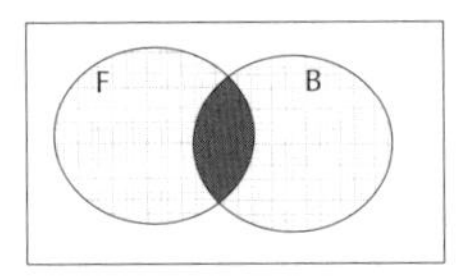

'exactly one 6' ∩ 'at least one 6' = 'exactly one 6'
(Look at the diagram — 'exactly one 6' is the cross-hatched area, and 'at least one 6' is the cross-hatched area plus the grey bit. So the bit in common to both is just the cross-hatched area.)
Now, that means $P(\text{exactly one 6} \cap \text{at least one 6}) = P(B \cap F') + P(B' \cap F)$
— this is the cross-hatched area in the Venn diagram,
i.e. $P(\text{exactly one 6} \cap \text{at least one 6}) = (1/5 \times 5/6) + (4/5 \times 1/6)$
$= 9/30 = 3/10$ *(using the fact that B and F are independent).*
$P(\text{at least one 6}) = 1/5 + 1/6 - (1/5 \times 1/6) = 10/30 = 1/3$
And all of this means $P(\text{exactly one 6} \mid \text{at least one 6}) = 3/10 \div 1/3 = 9/10$
[3 marks for the correct answer — otherwise up to 2 marks available for correct working]

Page 25

1) *STATISTICS has 10 letters, 3 repeated S's, 3 repeated T's and 2 repeated I's. This gives* $10! \div (3! \times 3! \times 2!) = 50400$ *arrangements.*

2) a) *the 6 people can sit in* $6! = 720$ *ways.*

b) *Put Mr & Mrs Brown together as one object — you can do this in two ways. This means there are 5 objects to rearrange (which can be done in 5! ways) and for each of these, there are 2 "Brown arrangements". This means there are* $2 \times 5! = 240$ *ways to seat the Browns together.*

c) *The number of ways the Browns sit apart is* $720 - 240 = 480$.
So $P(\text{Browns sit apart}) = 480 \div 720 = 2/3$

3) a) ${}^{26}C_6 = 26 \times 25 \times 24 \times 23 \times 22 \times 21 \div (6 \times 5 \times 4 \times 3 \times 2 \times 1) = 230230$ *ways*

b) *We can choose the 5 vowels in one way, leaving 21 choices for the last letter. This means there are 21 ways.*
So $P(5 \text{ vowels}) = 21 \div 230230 = 3/32890$

c) *Choosing no vowels is the complement of at least one vowel. No vowels can be chosen in* ${}^{21}C_6 = 54264$ *possible ways.*
$P(\text{no vowels}) = 54264 \div 230230 = 0.236$ *(to 3 sig. fig.)*
So $P(\text{at least 1 vowel}) = 1 - 0.236 = 0.764$ *(to 3 sig. fig.)*

d) *There are* $6 \times 5 \times 4 \times 3 = 360$ *ways to arrange 4 discs.*
You can start with a vowel in 3 ways, leaving 5 letters for 3 places so there are $3 \times 5 \times 4 \times 3 = 180$ *arrangements that start with a vowel.*
So $P(\text{starts with a vowel}) = 180/360 = 1/2$
(Or you could say that it just depends on the first tile, whether the arrangement starts with a vowel or not. There are 6 tiles, and 3 of these are vowels, so the probability of a random arrangement starting with a vowel is 1/2.)

4) a) *10 people can be arranged in* $10! = 3\,628\,800$ *ways. [1 mark]*

b) *You can stand the 5 women in* $5! = 120$ *ways.*
This leaves 6 gaps for the 5 men to stand in (including the places at the start and end of the line). If you number these gaps 1-6, then the men have to be standing in either gaps 1-5 or gaps 2-6 (otherwise two women stand together).
Ignoring where they stand for now, the men can be arranged in $5! = 120$ *ways. So for each of the 120 arrangements of women, there are* $2 \times 120 = 240$ *arrangements of the men (since they could be in gaps 1-5 or 2-6).*
This means the total number of ways of them all standing apart is $120 \times 240 = 28800$, *so the probability of this happening is* $28800 \div 3628800 = 1/126$
[3 marks for the correct answer — otherwise up to 2 marks available for correct working]

c) *You can have 2, 3, 4 or 5 men (corresponding to 4, 3, 2 or 1 women).*
Suppose 2 men and 4 women are to be chosen. There are ${}^5C_2 = 10$ *ways to choose 2 men, and* ${}^5C_4 = 5$ *ways to choose the 4 women. This means there would be* $10 \times 5 =$ *50 ways to choose the 2 men and 4 women.*
Similarly 3 men and 3 women could be chosen in ${}^5C_3 \times {}^5C_3 =$ *100 ways.*
4 men and 2 women could be chosen in ${}^5C_4 \times {}^5C_2 =$ *50 ways.*
5 men and 1 woman could be chosen in ${}^5C_5 \times {}^5C_1 =$ *5 ways.*
So the total number of ways in which the people could be chosen is:
$50 + 100 + 50 + 5 = 205$.
[3 marks for the correct answer — otherwise up to 2 marks available for correct working]

Section Three — Probability Distributions

Page 27

1) a) *All the probabilities have to add up to 1.*
So $0.5 + k + k + 3k = 0.5 + 5k = 1$, *i.e.* $5k = 0.5$, *i.e.* $k = 0.1$.

b) $P(Y < 2) = P(Y = 0) + P(Y = 1) = 0.5 + 0.1 = 0.6$.

2) *Make a table showing the possible values of X, i.e. total scores on the dice (though there are other ways to do this):*

Score on dice 1 (columns); Score on dice 2 (rows)

+	1	1	1	2	2	3
1	2	2	2	3	3	4
1	2	2	2	3	3	4
1	2	2	2	3	3	4
2	3	3	3	4	4	5
2	3	3	3	4	4	5
3	4	4	4	5	5	6

There are 36 outcomes in total, and of these 10 have X = 4.

So $P(X = 4) = \frac{10}{36} = \frac{5}{18}$.

Altogether there are 5 possible values for X, and the rest of the pdf is found in the same way that you found P(X = 4). The pdf is summarised in this table:

x	2	3	4	5	6
$P(X = x)$	$1/4$	$1/3$	$5/18$	$1/9$	$1/36$

3) a) *The probability of getting 3 heads is:* $\frac{1}{2} \times \frac{1}{2} \times \frac{1}{2} = \frac{1}{8}$ *[1 mark]*

The probability of getting 2 heads is: $3 \times \frac{1}{2} \times \frac{1}{2} \times \frac{1}{2} = \frac{3}{8}$ *(multiply by 3 because any of the three coins could be the tail — the order in which the heads and the tail occur isn't important). [1 mark]*

Similarly the probability of getting 1 head is: $3 \times \frac{1}{2} \times \frac{1}{2} \times \frac{1}{2} = \frac{3}{8}$. *[1 mark]*

And the probability of getting no heads is $\frac{1}{2} \times \frac{1}{2} \times \frac{1}{2} = \frac{1}{8}$. *[1 mark]*

Hence the pdf of X is:

x	20p	10p	nothing
$P(X = x)$	$1/8$	$3/8$	$1/2$

Answers

b) *You need the probability that X >10p [1 mark]. This is just*

P(X = 20p) = $\frac{1}{8}$ *[1 mark]*

Page 29

1) $P(W\le 0.2)=P(W=0.2)=0.2$

$P(W\le 0.3)=P(W=0.2)+P(W=0.3)=0.4$

$P(W\le 0.4)=P(W=0.2)+P(W=0.3)+P(W=0.4)=0.7$

$P(W\le 0.5)=P(W=0.2)+P(W=0.3)+P(W=0.4)+P(W=0.5)=1$

So the cumulative distribution function of W is:

w	0.2	0.3	0.4	0.5
P(W ≤ w)	0.2	0.4	0.7	1

2) $P(R=0)=P(R\le 0)=F(0)=0.1$

$P(R=1)=P(R\le 1)-P(R\le 0)=0.5-0.1=0.4$

$P(R=2)=P(R\le 2)-P(R\le 1)=1-0.5=0.5$

So the pdf of R is:

r	0	1	2
P(R = r)	0.1	0.4	0.5

$P(0\le R\le 1)=0.5$

3) *There are 5 possible outcomes, and the probability of each of them is k, so k = 1 ÷ 5 = 0.2.*

Mean of X = $\frac{0+4}{2}=2$.

Variance of X = $\frac{(4-0+1)^2-1}{12}=\frac{24}{12}=2$.

4) a) *All the probabilities must add up to 1, so 2k + 3k + k + k = 1, i.e. 7k = 1, and so* $k=\frac{1}{7}$. *[1 mark]*

b) $P(X\le 0)=P(X=0)=\frac{2}{7}$ *[1 mark]*

$P(X\le 1)=P(X=0)+P(X=1)=\frac{5}{7}$ *[1 mark]*

$P(X\le 2)=P(X=0)+P(X=1)+P(X=2)=\frac{6}{7}$ *[1 mark]*

$P(X\le 3)=P(X=0)+P(X=1)+P(X=2)+P(X=3)=1$ *[1 mark]*

So the distribution function is as in the following table:

x	0	1	2	3
P(X ≤ x)	2/7	5/7	6/7	1

c) $P(X>2)=1-P(X\le 2)=1-\frac{6}{7}=\frac{1}{7}$ *[1 mark]*

(Or P(X > 2) = P(X = 3) = $\frac{1}{7}$*, using part a).)*

5) a)

x	0	1	2	3	4	5	6	7	8	9
P(X = x)	0.1	0.1	0.1	0.1	0.1	0.1	0.1	0.1	0.1	0.1

[1 mark]

b) $\text{Mean}=\frac{0+9}{2}=4.5$ *[1 mark]* $\text{Variance}=\frac{(9-0+1)^2-1}{12}=\frac{99}{12}=8.25$ *[2 marks]*

c) $P(X<4.5)=P(X=0)+P(X=1)+P(X=2)+P(X=3)+P(X=4)=0.5$ *[2 marks]*

Page 31

1) a) *As always, the probabilities have to add up to 1, so*

$k=1-\left(\frac{1}{6}+\frac{1}{2}+\frac{5}{24}\right)=1-\frac{21}{24}=\frac{3}{24}=\frac{1}{8}$

b) $E(X)=\left(1\times\frac{1}{6}\right)+\left(2\times\frac{1}{2}\right)+\left(3\times\frac{1}{8}\right)+\left(4\times\frac{5}{24}\right)=\frac{4+24+9+20}{24}=\frac{57}{24}=\frac{19}{8}$

$E(X^2)=\left(1^1\times\frac{1}{6}\right)+\left(2^2\times\frac{1}{2}\right)+\left(3^2\times\frac{1}{8}\right)+\left(4^2\times\frac{5}{24}\right)=\frac{4+48+27+80}{24}=\frac{159}{24}$

$Var(X)=E(X^2)-\left[E(X)\right]^2=\frac{159}{24}-\left(\frac{57}{24}\right)^2=\frac{3816-3249}{576}=\frac{567}{576}=\frac{63}{64}$

c) $E(2X-1)=2E(X)-1=2\times\frac{19}{8}-1=\frac{30}{8}=\frac{15}{4}$

$Var(2X-1)=2^2Var(X)=4\times\frac{63}{64}=\frac{63}{16}$

2) a) *P(X = 1) = a, P(X = 2) = 2a, P(X = 3) = 3a. Therefore the total probability is 3a + 2a + a = 6a. This must equal 1, so* $a=\frac{1}{6}$. *[1 mark]*

b) $E(X)=\left(1\times\frac{1}{6}\right)+\left(2\times\frac{2}{6}\right)+\left(3\times\frac{3}{6}\right)=\frac{1+4+9}{6}=\frac{14}{6}=\frac{7}{3}$ *[2 marks]*

c) $E(X^2)=Var(X)+\left[E(X)\right]^2=\frac{5}{9}+\left(\frac{7}{3}\right)^2=\frac{5+49}{9}=\frac{54}{9}=6$ *[2 marks]*

d) $E(3X+4)=3E(X)+4=3\times\frac{7}{3}+4=11$ *[1 mark]*

$Var(3X+4)=3^2Var(X)=9\times\frac{5}{9}=5$ *[2 marks]*

Page 33

1) a) $P(X=2)=\binom{5}{2}0.3^2(1-0.3)^3=\frac{5!}{2!3!}\times 0.09\times 0.343=0.3087$

b) $P(X\le 3)=P(X=0)+P(X=1)+P(X=2)+P(X=3)$

$=\binom{5}{0}\times 0.3^0\times 0.7^5+\binom{5}{1}\times 0.3^1\times 0.7^4$

$+\binom{5}{2}\times 0.3^2\times 0.7^3+\binom{5}{3}\times 0.3^3\times 0.7^2$

$=0.96922$

c) $P(X<2)=P(X=0)+P(X=1)$

$=\binom{5}{0}\times 0.3^0\times 0.7^5+\binom{5}{1}\times 0.3^1\times 0.7^4$

$=0.52822$

d) $E(X)=np=5\times 0.3=1.5$

e) $Var(X)=npq=5\times 0.3\times(1-0.3)=1.05$

2) a) *This is a binomial distribution (call the random variable X) where n = 20 and p =* $\frac{1}{6}$ *('success' here is rolling a 6).*

$P(X=10)=\binom{20}{10}\times\left(\frac{1}{6}\right)^{10}\times\left(\frac{5}{6}\right)^{10}=4.93\times 10^{-4}(=0.000493)$ *to 3 sig. fig.*

(It's quickest to use tables for this really.)

b) *It's definitely quicker to use tables for this one:*

$P(X\ge 10)=1-P(X\le 9)=1-0.999401=0.000599$

3) *This is another binomial distribution, where X is 'the number of people eating salad', n = 10 and p = 0.3. You need P(X < 3).*

$P(X<3)=P(X=0)+P(X=1)+P(X=2)$

$=\binom{10}{0}\times 0.3^0\times 0.7^{10}+\binom{10}{1}\times 0.3^1\times 0.7^9+\binom{10}{2}\times 0.3^2\times 0.7^8$

$=0.3828$ *to 4 sig. fig.*

4) a) *This is a binomial distribution. Here, n = 30 and p =* $\frac{1}{7}$*, so if X is the number of people born on a Saturday,* $X\sim B\left(30,\frac{1}{7}\right)$. *[1 mark]*

The conditions needed for a binomial distribution are:
(i) there are a fixed number of trials,
(ii) there are just two possible outcomes of each trial,
(iii) each of the trials is independent of all the others,
(iv) the variable is the total number of successes.
[4 marks available — 1 mark for each condition]

Answers

b) *This is P(X = 5), and is given by:*

$$P(X=5)=\binom{30}{5}\left(\frac{1}{7}\right)^5\left(\frac{6}{7}\right)^{25}$$

$$=\frac{30!}{5!25!}\left(\frac{1}{7}\right)^5\left(\frac{6}{7}\right)^{25}=0.1798 \text{ to 4 sig. fig.} \quad [2 \text{ marks}]$$

Page 35

1) $P(X=15)=0.9^{14}\times 0.1=0.0229$ to 3 sig. fig.

The expected value of X is 1 ÷ 0.1 = 10.

2) a) *When you throw two dice, there are 36 possible outcomes and 6 of these will be doubles. So if X is the random variable 'number of throws till you get a double', then X ~ Geo(* $\frac{1}{6}$ *). [1 mark]*

Then the mean number of throws needed to start the game is 1 ÷ $\frac{1}{6}$ *= 6. [1 mark]*

b) (i) $P(X=4)=\left(\frac{5}{6}\right)^3\times\frac{1}{6}=0.0965$ *[2 marks]*

(ii) *If it takes at least 5 throws, then that means that the first four throws were unsuccessful [1 mark]. Since the probability of 'not a success' is* $\frac{5}{6}$*, then the required probability must be* $\left(\frac{5}{6}\right)^4=0.482$*. [2 marks]*

c) *The probability that a player has started by the time he/she's had four throws is* $1-\left(\frac{5}{6}\right)^4$ *(using the above). Since all throws are independent, the probability that both players have started is* $\left[1-\left(\frac{5}{6}\right)^4\right]^2$ *[1 mark]. So the probability that at least one of them hasn't started is*

$$1-\left[1-\left(\frac{5}{6}\right)^4\right]^2=0.732 \text{ to 3 sig. fig.} \quad [1 \text{ mark}]$$

Page 37

1) a) *Normalise the probabilities and then use tables to answer these. Remember, X ~ N(50, 16) (and 16 is the variance, not the standard deviation).*

$$P(X<55)=P\left(Z<\frac{55-50}{\sqrt{16}}\right)=P(Z<1.25)=0.8944$$

b) $P(X<42)=P\left(Z<\frac{42-50}{\sqrt{16}}\right)=P(Z<-2)$

$$=1-P(Z<2)=1-0.9772=0.022$$

c) $P(X>56)=1-P(X<56)=1-P\left(Z<\frac{56-50}{\sqrt{16}}\right)=$

$$1-P(Z<1.5)=1-0.9332=0.0668$$

d) $P(47<X<57)=P(X<57)-P(X<47)=P\left(Z<\frac{57-50}{\sqrt{16}}\right)-P\left(Z<\frac{47-50}{\sqrt{16}}\right)$

$$=P(Z<1.75)-P(Z<-0.75)$$
$$=P(Z<1.75)-(1-P(Z<0.75))$$
$$=0.9599-(1-0.7734)$$
$$=0.7333$$

2) a) *Here X ~ N(600, 202). You need to use your 'percentage points' table for these.*

If $P(X<a)=0.95$, then $P\left(Z<\frac{a-600}{\sqrt{202}}\right)=0.95$.

So $\frac{a-600}{\sqrt{202}}=1.645$ (using the table).

Rearrange this to get $a=600+1.645\times\sqrt{202}=623.3$

b) *|X – 600| < b means that X is 'within b' of 600, i.e. 600 – b < X < 600 + b. Since 600 is the mean of X, and since a normal distribution is symmetrical,*

$P(600-b<X<600+b)=0.8$ means that $P(600<X<600+b)=0.4$

i.e. $P\left(\frac{600-600}{\sqrt{202}}<Z<\frac{600+b-600}{\sqrt{202}}\right)=0.4$, i.e. $P\left(0<Z<\frac{b}{\sqrt{202}}\right)=0.4$

This means that $P\left(Z<\frac{b}{\sqrt{202}}\right)-P(Z<0)=0.4$

i.e. $P\left(Z<\frac{b}{\sqrt{202}}\right)-0.5=0.4$, or $P\left(Z<\frac{b}{\sqrt{202}}\right)=0.9$

Use your percentage points table to find that

$$\frac{b}{\sqrt{202}}=1.2816, \text{ or } b=1.2816\times\sqrt{202}=18.21$$

3 a) *Here, the random variable X (the distribution of the marks) is distributed X ~ N(50, 30²).*
First, you need P(X > 41) — this will tell you the fraction of marks that are above 41.

$$P(X>41)=P\left(Z>\frac{41-50}{30}\right)=P(Z>-0.3)$$

This is equal to $P(Z<0.3)=0.6179$ (from tables). *[2 marks]*

Then to estimate the number of candidates that passed the exam, multiply this by 1000 — so roughly 618 students are likely to have passed [1 mark].

b) *Let the mark required for an A-grade be k. Then since 90% of students don't get an A, P(X < k) = 0.9 [1 mark]*

Normalise this to get $P\left(Z<\frac{k-50}{30}\right)=0.9$ *[1 mark]*

Now you can use your percentage points table to get that

$\frac{k-50}{30}=1.282$, or $k=30\times1.282+50=88.45$ *[1 mark]. So the mark needed for an A-grade will be around 88-89 marks.*

4) *Assume that the lives of the batteries are distributed as* $N(\mu,\sigma^2)$.

Then P(X < 20) = 0.4 [1 mark] and P(X < 30) = 0.8.
Normalise these 2 equations to get

$$P\left(Z<\frac{20-\mu}{\sigma}\right)=0.4 \text{ and } P\left(Z<\frac{30-\mu}{\sigma}\right)=0.8 \quad [1 \text{ mark}]$$

Now you need to use your percentage points table to get:

$$\frac{20-\mu}{\sigma}=-0.2533 \text{ and } \frac{30-\mu}{\sigma}=0.8416 \quad [2 \text{ marks}]$$

Now rewrite these as:

$20-\mu=-0.2533\sigma$ and $30-\mu=0.8416\sigma$. *[2 marks]*

Subtract these two equations to get:

$$10=(0.8416+0.2533)\sigma$$

i.e. $\sigma=\frac{10}{0.8416+0.2533}=9.1333$ *[1 mark]*

Now use this value of σ in one of the equations above:

$\mu=20+0.2533\times9.1333=22.31$ *[1 mark]*

So $X \sim N(22.31, 9.13^2)$ i.e. $X \sim N(22.31, 83.4)$

Page 39

1) *This is a binomial distribution, but we can use the normal approximation here, since n is very large (or np = 10 and nq = 190, both of which are bigger than 5).*
To find the correct normal distribution, we need the mean (= np). This is 200 × 0.05 = 10. The variance (= npq) is given by 200 × 0.05 × 0.95 = 9.5.
So we need to use the approximation A ~ N(10, 9.5).
We need P(A < 11) — using a continuity correction means you have to actually find P(A < 10.5).

Answers

Now you can normalise as usual...

$$P(A<10.5)=P\left(Z<\frac{10.5-10}{\sqrt{9.5}}\right)=P(Z<0.1622)$$

Using tables this is 0.5644

So the probability that A is less than 11 is 0.5644.

2) *This is another binomial question — here, if X is the number of plants that are short of water, X ~ B(500, 0.3).*
Again, we can use a normal approximation. This is justified, since n is so large (or np = 150 and nq = 350 are both very large).
The mean (= np) is 500 × 0.3 = 150.
The variance = npq = 500 × 0.3 × 0.7 = 105.
So approximately, X ~ N(150, 105).
We need P(X < 149), which, with the continuity correction, means finding P(X < 148.5). Now you can normalise as usual...

$$P(X<148.5)=P\left(Z<\frac{148.5-150}{\sqrt{105}}\right)=P(Z<-0.1464)=1-P(Z<0.1464)$$

Using tables, this is $1-0.5582=0.4418$

3) *The probability of getting 2 sixes with two throws is* $\frac{1}{6}\times\frac{1}{6}=\frac{1}{36}$ *[1 mark]. So if X is the number of people who get 2 sixes,* $X\sim B\left(200,\frac{1}{36}\right)$ *[1 mark].*
Once again, you can use a normal approximation, since n is so large (or because $200\times\frac{1}{36}=5.6$ *and* $200\times\frac{35}{36}=194.4$ *are both bigger than 5) [1 mark]. The mean of X is np = 5.556, and the variance is*
$200\times\frac{1}{36}\times\frac{35}{36}=5.401$.
So the approximation you need is X ~ N(5.556, 5.401) [1 mark].
You need $P(X\geq 10)$, *so with the continuity correction this is*
$P(X>9.5)=1-P(X<9.5)$ *[1 mark].*
Now normalise as usual...

$$1-P(X<9.5)=1-P\left(Z<\frac{9.5-5.556}{\sqrt{5.401}}\right)=1-P(Z<1.697)$$

Using tables, this is $1-0.9552=0.0448$ *[1 mark]*

4) a) *If X is the random variable distributed as the weight of the bags of sweets, then X ~ N(30, 5²) = N(30, 25), and so for a sample of 50 bags, the mean is distributed as* $\bar{X}\sim N\left(30,\frac{25}{50}\right)=N(30,0.5)$. *[1 mark]*

b) *Once again, just normalise...*

$$P(\bar{X}<29)=P\left(Z<\frac{29-30}{\sqrt{0.5}}\right)=P(Z<-1.414)$$
$$=1-P(Z<1.414)=1-0.9213=0.0787 \text{ [2 marks]}$$

c) *This time, we have a sample of size n, so the distribution of the sample mean is* $\bar{X}\sim N\left(30,\frac{25}{n}\right)$ *[1 mark]. We need n big enough that*
$P(\bar{X}<29)=0.05$ *[1 mark]*
Normalise again...

$$P(\bar{X}<29)=P\left(Z<\frac{29-30}{\sqrt{25/n}}\right)=0.05$$

Use your tables to find that $\frac{29-30}{\sqrt{25/n}}=-1.645$ *[1 mark]*

So $\sqrt{\frac{25}{n}}=0.6079$, *and* $\frac{25}{n}=0.3695$,
which means that $n=67.7$ *[1 mark]*

So if we choose n to be 68 or more, we should be all right [1 mark].

Section Four — Correlation and Regression

Page 41

1)

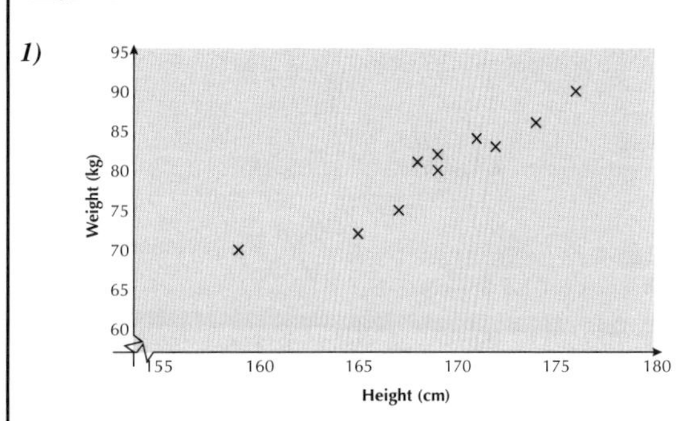

I've called the height x and the weight y.
You need to work out these sums:

$\sum x=1690$, $\sum y=803$, $\sum x^2=285818$, $\sum y^2=64835$, $\sum xy=135965$.

$$\text{Then } r=\frac{135965-\frac{1690\times 803}{10}}{\sqrt{\left(285818-\frac{1690^2}{10}\right)\left(64835-\frac{803^2}{10}\right)}}=\frac{258}{\sqrt{208\times 354.1}}=0.951$$

This value is very close to 1, which tells you that the variables 'height' and 'weight' have a high positive correlation — as the height increases, generally the weight increases as well.

2)

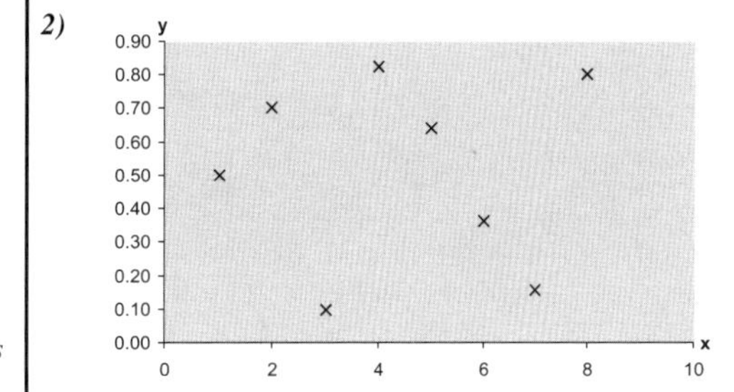

[Up to 2 marks available]
Again, you need to work out these sums:

$\sum x=36$, $\sum y=4.08$, $\sum x^2=204$, $\sum y^2=2.6272$, $\sum xy=18.36$.
[5 marks available — 1 for each correct sum]

$$\text{Then } r=\frac{18.36-\frac{36\times 4.08}{8}}{\sqrt{\left(204-\frac{36^2}{8}\right)\left(2.6272-\frac{4.08^2}{8}\right)}}=\frac{0}{\sqrt{42\times 0.5464}}=0 \text{ [1 mark]}$$

This value of zero for the correlation tells you that there appears to be no linear relationship between the two variables [1 mark].

Page 43

1) *It's best to draw a table like the one below:*

Physics	54	34	23	58	52	58	13	65	69	52
English	16	73	89	81	23	81	56	62	61	37
Physics rank	5	8	9	3.5	6.5	3.5	10	2	1	6.5
English rank	10	4	1	2.5	9	2.5	7	5	6	8
d	-5	4	8	1	-2.5	1	3	-3	-5	-1.5
d²	25	16	64	1	6.25	1	9	9	25	2.25

$\sum d^2=158.5$

$$\text{Then } r_s=1-\frac{6\sum d^2}{n(n^2-1)}=1-\frac{6\times 158.5}{10\times(10^2-1)}=1-\frac{951}{990}=0.0394 \text{ to 3 sig. fig.}$$

This is quite low, so it appears that doing well in Physics is no indication as to how well you will do in English and vice versa.

Answers

2) a) *First work out S_{xy}, S_{xx} and S_{yy}:*

$$S_{xy} = 26161 - \frac{386 \times 460}{8} = 3966 \text{ [1 mark]}$$

$$S_{xx} = 25426 - \frac{386^2}{8} = 6801.5 \text{ [1 mark]}$$

$$S_{yy} = 28867 - \frac{460^2}{8} = 2417 \text{ [1 mark]}$$

Then $r = \frac{S_{xy}}{\sqrt{S_{xx}S_{yy}}} = \frac{3966}{\sqrt{6801.5 \times 2417}} = 0.978$ to 3 sig. fig. *[2 marks]*

b) *Since r is very close to 1, the quantities x and y are very closely positively correlated. This means that the more money spent on advertising one of the products, the higher the sales tend to be. [2 marks]*

Page 45

1) a)

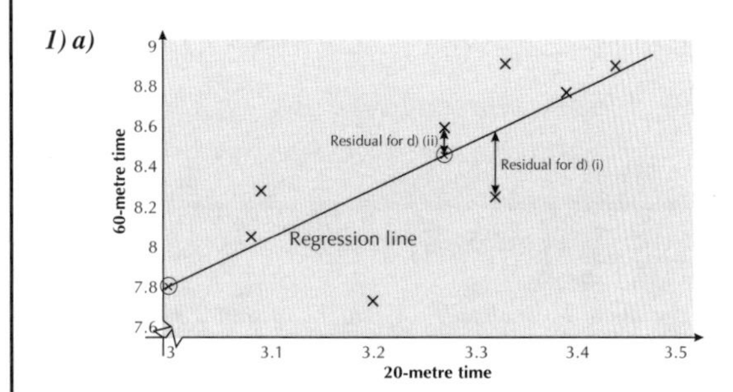

[3 marks available for the scatter diagram]

b) *It's best to make a table like this one, first:*

									Totals
20-metre time, x	3.39	3.2	3.09	3.32	3.33	3.27	3.44	3.08	26.12
60-metre time, y	8.78	7.73	8.28	8.25	8.91	8.59	8.9	8.05	67.49
x^2	11.4921	10.24	9.5481	11.0224	11.0889	10.6929	11.8336	9.4864	85.4044
y^2	77.0884	59.7529	68.5584	68.0625	79.3881	73.7881	79.21	64.8025	570.6509
xy	29.7642	24.736	25.5852	27.39	29.6703	28.0893	30.616	24.794	220.645

[2 marks available for at least 3 correct totals on the right-hand side]

Then: $S_{xy} = 220.645 - \frac{26.12 \times 67.49}{8} = 0.29015$ *[1 mark]*

$$S_{xx} = 85.4044 - \frac{26.12^2}{8} = 0.1226 \text{ [1 mark]}$$

Then the gradient b is given by: $b = \frac{S_{xy}}{S_{xx}} = \frac{0.29015}{0.1226} = 2.3666$ *[1 mark]*

And the intercept a is given by:

$$a = \bar{y} - b\bar{x} = \frac{\sum y}{n} - b\frac{\sum x}{n} = \frac{67.49}{8} - 2.3666 \times \frac{26.12}{8} = 0.709 \text{ [1 mark]}$$

So the regression line has equation: y = 2.367x + 0.709 [1 mark]
To plot the line, find two points that the line passes through. A regression line always passes through $(\bar{x}, \bar{y})$, *which here is equal to (3.27, 8.44).*

Then put x = 3 (say) to find that the line also passes through (3, 7.81). Now plot these points (in circles) on your scatter diagram, and draw the regression line through them [1 mark for plotting the line correctly]

c) ***(i)*** *x = 3.15, y = 2.367 × 3.15 + 0.709 = 8.17 (to 3 sig. fig.) [2 marks] — this should be reliable, since we are using interpolation in a known region [1 mark].*
(i) *x = 3.88, y = 2.367 × 3.88 + 0.709 = 9.89 (to 3 sig. fig.) [2 marks] — this could be unreliable, since we are using extrapolation [1 mark].*

d) ***(i)*** *x = 3.32, residual = 8.25 – (2.367 × 3.32 + 0.709) = –0.317 (3 sig. fig.) [1 mark for calculation, 1 mark for plotting residual correctly]*
(ii) *x = 3.27, residual = 8.59 – (2.367 × 3.27 + 0.709) = 0.141 (3 sig. fig.) [1 mark for calculation, 1 mark for plotting residual correctly]*

Section Five — Hypothesis Testing

Page 47

1) a) $H_0 : p = 0.35$
$H_1 : p \neq 0.35$
This leads to a two-tail test.

b) $H_0 : p = 0.5$
$H_1 : p > 0.5$
This is a one-tail test.

c) $H_0 : p = 0.1$
$H_1 : p < 0.1$
This is a one-tail test.

d) $H_0 : p = 0.4$ *(this is the null hypothesis of 'no change')*
$H_1 : p > 0.4$
This is a one-tail test.

e) $H_0 : p = 0.2$
$H_1 : p < 0.2$
This is a one-tail test.

Page 49

1) a) *Critical Region = {3 or fewer "successes")*
since P(3 or fewer) = 0.0160 < 0.05
but P(4 or fewer) = 0.0510 > 0.05

b) *Critical Region = {10 or more "successes"}*
since P(10 or more) = 1 – P(9 or fewer) = 1 – 0.9520 = 0.0480 < 0.05
but P(9 or more) = 1 – P(8 or fewer) = 1 – 0.8867 = 0.1133 > 0.05

c) *Find the lower and upper critical regions separately — and use 'half the total significance level' for each bit.*
First the lower critical region — this is {0 "successes"},
since P(1 or fewer) = 0.0692 > 0.025
but P(0) = 0.0115 < 0.025
Then the upper critical region — this is {9 or more "successes"},
since P(9 or more) = 1 – P(8 or fewer) = 1 – 0.9900 = 0.0100 < 0.025
but P(8 or more) = 1 – P(7 or fewer) = 1 – 0.9679 = 0.0321 > 0.025
So the overall critical region = {0 "successes" or *9 or more "successes"}*

Page 51

1) a) $p = \frac{300}{500} = 0.6$

This is the probability of a "success", i.e. that a person says they intend to vote for the People's Party at the next election [1 mark].

b) $H_0 : p = 0.6$ *(This is the null hypothesis of 'no change'.) [1 mark]*
$H_1 : p > 0.6$ *(You want to reject the 'no change' null hypothesis if p has risen.) [1 mark]*

c) *This is a one-tail test, at a significance level of 5%.*
Critical Region = {17 or more "successes"} [1 mark]
since P(17 or more) = 1 – P(16 or fewer) = 1 – 0.9840 = 0.0160 < 0.05 [1 mark]
but P(16 or more) = 1 – P(15 or fewer) = 1 – 0.9490 = 0.0510 > 0.05 [1 mark]

d) *Test statistic = 17 "successes" observed (i.e. 17 people in the sample said they intended to vote for the "people's party" in the next election).*
The test statistic is in the critical region so reject H_0 [1 mark].
Conclude that there is evidence to support the campaign manager's claim [1 mark].

Index

Index